Nuclear Pursuits

Nuclear Pursuits

The Scientific Biography of Wilfrid Bennett Lewis

RUTH FAWCETT

McGill-Queen's University Press
Montreal & Kingston • London • Buffalo

ISBN 0-7735-1186-5

Legal deposit second quarter 1994
Bibliothèque nationale du Québec

Printed in Canada on acid-free paper

This book has been published by the Social Science Federation of Canada, using funds provided by the Social Sciences and Humanities Research Council of Canada. Publication has also been supported by the Canada Council through its block grant program.

Canadian Cataloguing in Publication Data

Fawcett, Ruth
Nuclear pursuits: the scientific biography of Wilfrid Bennett Lewis
Includes bibliographical references and index.
ISBN 0-7735-1186-5
1. Lewis, W. Bennett (Wilfrid Bennett), 1908–1987.
2. Physicists – Canada – Biography. 3. Nuclear energy – Research – Canada. I. Title.
QC774.L49F39 1994 539.7'092 C94-900062-0

Illustrations are courtesy Atomic Energy of Canada Limited unless otherwise indicated.

This book was typeset by Typo Litho Composition Inc. in 10/12 Palatino.

Contents

Acknowledgments

The research and writing of a book involves the help and support of a great many people.

This book grew out of a thesis prepared under the supervision of Professor Bob Bothwell at the University of Toronto. He encouraged me in this endeavour from the beginning and his help throughout has been invaluable. I am also grateful to Professors Craig Broun and Michael Bliss for many helpful comments.

Atomic Energy of Canada Limited was extremely generous in allowing me free and open access to all their files pertaining to W.B. Lewis. My research was largely carried out at the Chalk River Nuclear Laboratories (CRNL), where Ruth Everson and Carol Lachapelle were enormously helpful and patient in guiding me to a vast number of files. I cannot thank them enough. Morris Dament happily retrieved a seemingly endless number of lectures, memoranda, and reports by Lewis stored at the Scientific Documents Distribution Office. I also appreciate the help given by Ralph Robinson, Phyllis Crozier, and Gwynedd Gerry for help with the photographs.

Staff at the National Archives of Canada, the Public Records Office in London, and the Churchill College Archives Centre in Cambridge were very helpful. Janet Dudley at the Royal Signals and Radar Establishment directed me to Telecommunications Research Establishment (TRE) documents still stored in Malvern.

A great number of scientists and engineers from both Chalk River and Ontario Hydro helped me to gain a clearer picture of Lewis's abilities and personality through interviews. I am particularly grateful to Geoff Hanna for his tireless efforts, his patience, and most of all, his encouragement and support.

My editors, Claire Gigantes and Joan McGilvray, have been thorough with my manuscript and patient with me. I thank them for their support.

Financial assistance from the Social Sciences and Humanities Research Council of Canada, the Associates of the University of Toronto, the Centre for International Studies, the School of Graduate Studies, and the University of Toronto is gratefully acknowledged.

A great many friends have offered encouragement and the opportunity to relax over the past five years. Thanks particularly to Kay and George Rich for a lovely stay in Malvern, the lunch-room gang at Chalk River, and Mary Halloran for well-timed pep talks. My colleagues at the Science and Technology Division of the Library of Parliament's Research Branch have been enthusiastic about my extra-curricular activities. I am grateful also to Roger Sarty for helpful discussions and Glenn and Joan Sarty for Cape Cod rest and relaxation.

My husband, Leigh Sarty, has been wonderful throughout. He now knows more about W.B. Lewis and nuclear power reactors than I'm sure he ever believed possible. Without him, this book could not have been written. My son, Michael, has not yet been given any editorial duties but his cheery presence is a constant joy. My parents, Pat and Eric, have been understanding and supportive throughout my metamorphosis from physicist to historian. This book is dedicated to them.

Introduction

This is a study of the career of Wilfrid Bennett Lewis, the physicist who dominated nuclear research and the development of nuclear power in Canada for nearly three decades, from the end of World War II until his retirement in 1973.

Throughout this period and up to the present day, atomic energy has been a leading sector for government scientific research and development in Canada. The impact – on federal and provincial policy, universities, industry, Canada's place in international science, and her foreign relations – has been enormous. Indeed, the very scope of the atomic energy program and the intense political controversy it arouses have been a major impetus in the formation of a federal-government science policy.

Wilfrid Bennett Lewis was at the centre of these developments for twenty-seven years. As scientific leader at the Chalk River Nuclear Laboratories, Lewis played a vital role in the Canadian development of nuclear power. But despite his dominant role in Canada's scientific history his name is known today only to specialized scientists and engineers.

This is not atypical for the history of science in Canada, which has only recently started to receive the attention it deserves. In many ways this is a reflection of the lack of regard generally afforded science in Canada. Despite many significant scientific achievements in their country's past, Canadians still view themselves essentially as "hewers of wood and drawers of water." And with their typical humility they are hesitant to celebrate the obvious successes Canada has fostered.

The problem is magnified by the fact that Lewis worked in a field of science difficult for the layman to understand. More importantly, it is a field that, in recent years, has come to be regarded with suspicion and fear. Such attitudes, fuelled and seemingly confirmed by the dramatic accidents at Three Mile Island and Chernobyl, have strongly curbed the enthusiasm that surrounded nuclear power in its heyday.

Lewis arrived in Canada to head the nation's fledgling atomic energy establishment at the beginning of a period of growing belief in the promise of nuclear energy. His own faith in the powers of the atom matched the confidence prevalent throughout Canada as the postwar years began. Proud of the country's war effort, Canadians were ready to enjoy the prosperity of the postwar boom. In the field of atomic energy they had particular reason to be pleased. Wartime agreements had led to Canada's intimate involvement in one of the most significant scientific developments of the twentieth century. The challenge of the years after 1945 would be to encourage further progress in this area.

Lewis met this challenge head on. While maintaining an active research effort at Chalk River he led scientists and engineers in the development of the CANDU reactor. His philosophical and religious beliefs reinforced his conviction that nuclear power had a central role to play in humankind's progress and that Canada should be an active participant in its development. The scientific and technological achievements of Chalk River increased Canada's international scientific stature. Lewis became the nation's "nuclear statesman" representing the country around the world. Within Canada he was well known in the scientific community for his single-minded pursuit of scientific goals – from the research and demonstration reactors to the CANDU reactor and, later, the Intense Neutron Generator (ING).

By the end of his career, however, the wheel had turned and nuclear power was no longer in public favour. Lewis, unable or unwilling to understand the views of those who opposed nuclear power, continued to argue its benefits long into his retirement years. For Lewis, the optimism surrounding nuclear power that had built the industry (and with it, Lewis's career) in the 1950s remained an article of faith long after the promise had faded under the pressure of economic, environmental, and other social concerns.

In a sense, Lewis can be seen as simply a product of his times, part of a generation that retained its faith in the ability of science to improve the quality of life. But given his enormous individual impact on Canada's nuclear program, it is reasonable to ask whether his dogmatic approach contributed to the polarization of views that has characterized the nuclear debate. A definitive answer to this question is clearly unobtainable, but an examination of Lewis's career brings greater understanding of a subject that remains controversial to this day.

This biography deals largely with Lewis's scientific career. He was a very private man, wholly dedicated to his job and with only a couple of hobbies. No diary and few personal letters exist to give clues about his inner thoughts. But by conducting interviews with his colleagues and closely examining his writings, it has been possible to acquire a sense of Lewis's commanding personality and powerful intellect, which, it is hoped, is communicated in the following pages. For these were the qualities that enabled Lewis to play a decisive role both in the development of Canada's nuclear power reactors and the establishment of a tradition of excellence in fundamental nuclear research.

Abbreviations

AB	United Kingdom Atomic Energy Authority Records
AECB	Atomic Energy Control Board
AECL	Atomic Energy of Canada Limited
AVIA	Ministry of Aircraft Production Records
BEPO	British Experimental Pile Operation
CANDU	Canadian Deuterium Uranium
CIR	Canada-India Reactor
CRNL	Chalk River Nuclear Laboratories
DL	Director's Lectures
DM	Director's Memoranda
DR	Director's Reports
DSIR	Department of Scientific and Industrial Research
FO	Foreign Office Records
IAEA	International Atomic Energy Agency
ING	Intense Neutron Generator
NA	National Archives of Canada
NPD	Nuclear Power Demonstration
NPPD	Nuclear Power Plant Division
NRC	National Research Council
NRU	National Research Universal
NRX	National Research Experimental
PRDPEC	Power Reactor Development Programme Evaluation Committee
PRO	Public Records Office
PTR	Pool Test Reactor
RSRE	Royal Signals and Radar Establishment
SDDO	Scientific Documents Distribution Office

TRE	Telecommunications Research Establishment
UNSAC	United Nations Scientific Advisory Committee
WBL	Wilfrid Bennett Lewis
ZEEP	Zero Energy Experimental Pile

Wilfrid Bennett Lewis in December 1914 with his older sister Betty, his younger sister Gwynedd, and family pet Caesar. (Photograph courtesy of Mrs G. Gerry)

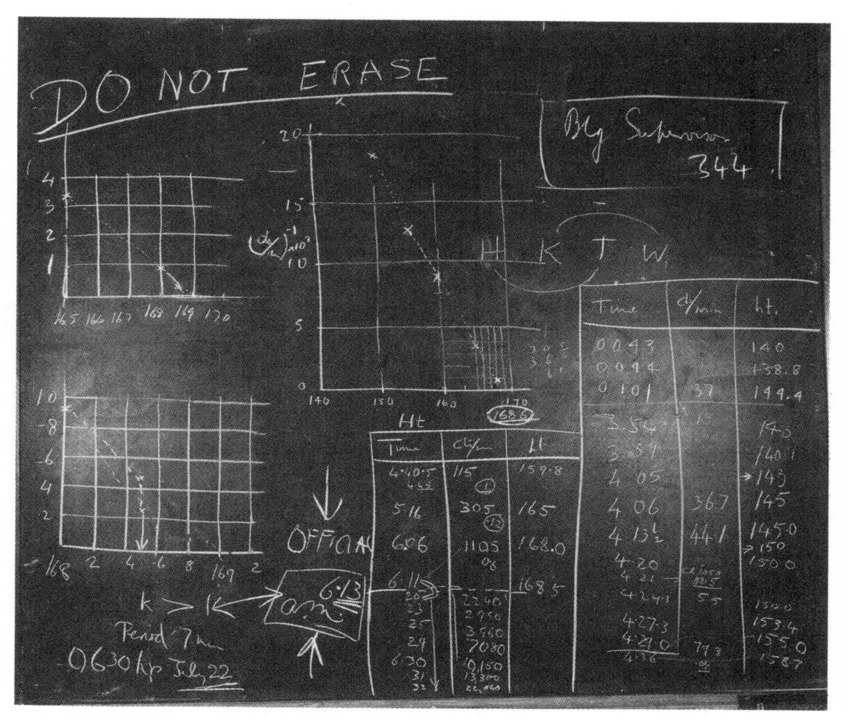

Blackboard showing calculation of NRX's start-up.

Professors and students at the Cavendish, June 1932 N.S. Alexander, P. Wright, A.G. Hill, J.L. Pawsey, G. Occhialina, H. Miller, W.E. Duncanson, E.C. Childs, T.G.P. Tarrant, J. McDougall, R.C. Evans, E.S. Shire, E.L.C. White, F.H. Nicoll, R.M. Chaudhri, B.V. Bowden, W.B. Lewis, P.C. Ho, C.B. Mohr, H.W.S.Massey, M.L. Oliphant, E.T.S. Walton, C.E. Wynn-Williams, J.K. Roberts, N. Feather, Miss Davies, Miss Sparshott, J.P. Gott, J.A. Ratcliffe, P. Kapitza, J. Chadwick, R. Ladenberg, Prof. Sir J.J. Thomson, Prof. Lord Rutherford, Prof. C.T.R. Wilson, F.W. Aston, C.D. Ellis, P.M.S. Blackett, J.D. Cockcroft

A group of Cavendish researchers gathered around the Rutherford experimental magnet used for the deflection of alpha particles. *Left to right*: C.B.O. Mohr, A.I. Leipunsky, M. Goldhaber, G.E. Pringle, P.I. Dee, Berthwhistle, C.W. Gilbert, W.B. Lewis, H. Carmichael, W.E. Bennett.

Lewis with senior scientists at the Telecommunications Research Establishment in 1945. *Left to right*: J.A. Ratcliffe, R.A. Smith (standing), C. Holt Smith, R. Cockburn (standing), A.P. Rowe, W.B. Lewis. (Crown Copyright, Defence Research Agency, Malvern)

The Changing of the Guard – September 1946 The Byeways Hotel was the scene of this historic gathering to bid farewell to J.D. Cockcroft and welcome W.B. Lewis. *Standing left to right*: A.J. Cipriani, B.W. Sargent, E.J. Wiggins, J.D. Cockcroft, L.G. Cook, G.C. Butler, T.W. Boyer, H.J. Desbarats, H. Tongue, K.F. Tupper, D.G. Hurst, A.M. Budden, D.C. Jones, W.B. Lewis, P.R. Wallace, P. Demers. *Seated left to right*: E.H. Denike, J.A. Cruikshank, E.P. Hincks, I.D. Ritchie, G.M. Volkoff, H.J. Clayton, L.G. Elliott, J.D. Stewart.

Lewis (second from left) with the model of NRX which was displayed at the 1955 United Nations Conference on the Peaceful Uses of Atomic Energy held in Geneva.

In May 1953 the highly radioactive NRX calandria is towed to its burial spot after the 1952 accident.

Experimental apparatus positioned around the NRX reactor in 1958.

Construction of the Canada-India reactor at Trombay, India, in 1958.

Aerial view of the Chalk River plant in 1958.

NPD (Nuclear Power Demonstration).

Aerial view of the town of Deep River.

Lewis with senior scientists in front of the EN Tandem accelerator in 1964. *Left to right*: A.E. Litherland, W.B. Lewis, R.J. Van de Graaff, L.G. Elliott, E. Almqvist, R.B. Tomlinson.

Nuclear Pursuits

CHAPTER ONE

Early Years

Wilfrid Bennett Lewis was born on 24 June 1908 in Castle Carrock, Cumberland, a northern county of England. His father, Arthur Wilfrid Lewis, had been posted as a resident engineer on the Carlisle waterworks at Castle Carrock. There he met and married Isoline Maud Steavenson, daughter of a county-court judge. They remained in Castle Carrock until after the birth of their second child, Wilfrid Bennett (their eldest child, Elizabeth Fenwick, had been born three years earlier), moving then to Beckenham in Kent. There, two more children were born, completing the family: a son, John Addison, and a daughter, Gwynedd Emily.[1]

A strong engineering tradition existed in the Lewis family. Its origins can be traced to Wilfrid Bennett's great-grandfather, a well-established tailor in Worcester, whose interests also ranged into politics. It is believed that while acting as alderman he became interested in railway promotion, which helped to guide his two sons, William Bourne and Edward James, into engineering. William Bourne (Wilfrid Bennett's paternal grandfather) trained with an engineer working under the famous engineer Isambard Kingdom Brunel on the Oxford, Worcester, and Wolverhampton Railway. After his apprenticeship William Bourne accepted a position in Brunel's office and continued to work on railways and other projects.[2]

Two of William Bourne Lewis's four sons became engineers; both trained with Messrs J. Mansergh and Sons on waterworks. It was this that brought one of them, Arthur Wilfrid Lewis, to Castle Carrock. With that project completed, the family moved south at the end of 1909 and by May 1910 Arthur Wilfrid Lewis had joined with two other engineers to form a partnership for the construction of lighthouses. The death of his partners in 1917 left Lewis in a position to invite his brother to join him in an alliance that lasted until their deaths during the Second World War. They worked largely on reservoirs, water supplies, and sea defences.

With his father, uncle, and paternal grandfather all employed in engineering, it is perhaps not surprising that Wilfrid Bennett Lewis (Ben to family and friends) showed early signs of mechanical ability. At a young age, his mother recollected, he had a box in which he kept his "electrical apparatus." Meccano was another favourite toy, but Lewis later recalled refusing to follow the instruction booklet, preferring instead to design his own models. His siblings remember his immense powers of concentration as he lay on the playroom floor, immersed in a book, while the others played around and over him. As he grew older, however, the hobby that occupied most of his free time was the study of wireless, or, as it is known today, radio.

Wireless telegraphy was invented around the turn of the century; shortly afterwards, further developments made it possible to transmit speech over the air. Rapid improvements allowed both the army and navy to use the new technology during the First World War, but it was only after hostilities had ceased and the ban on transmissions lifted that wireless became popular with the general public. Because the components used in the construction of transmitters and receivers were relatively uncomplicated, the field was perfect for amateurs with a technical bent. This description suited Wilfrid Bennett Lewis, who, as a teenager after the war, spent most of his free time constructing wireless sets. One whole side of his bedroom was converted into a work bench. On it he gathered together the bits and pieces needed to construct his apparatus. Using parts gathered from various stores in the East End of London and with the guidance of *Wireless World*, a magazine devoted to the subject, he built a wireless. Then, once the correct spot on the crystal had been located to tune in a station, the whole family would gather round to listen.[3]

From 1916 until 1922 Lewis attended the Clare House Preparatory School in Beckenham, Kent. He then moved on to the English public school (private, in North American terms) Haileybury College in Hertford. Academic distinction came only in his final year when he won the Senior Chemistry and Senior Physics prizes.[4] But his passion for wireless continued unabated. As he recalled later, "from 1922 onwards I had built up at home a radio laboratory on a small scale, making my own measuring apparatus, moving coil galvanometers, standard cell, potentiometer bridges as well as the radio apparatus."[5] On the basis of this experimental work at home Lewis later published a number of articles in the journal *Wireless Engineer and Experimental Wireless*.[6] His enthusiasm for wireless did at times interfere with his work at Haileybury College. On one occasion, his housemaster found that Lewis had been failing to complete his preparatory work because he had been writing an article for *Wireless World*.[7]

Lewis graduated from Haileybury College in 1926 and found a job at the research laboratory of Chance Bros and Co. Ltd, where he helped with work on optical glass. In 1927 he went up to Cambridge as a commoner (i.e., without a scholarship), relying on his father's support throughout his undergraduate years.[8]

Given Lewis's profound interest in electronics and physics it is not surprising that he chose to go to Cambridge. It was home to the famous Cavendish laboratory, known throughout the scientific world as a leading centre for experimental physics. Lewis would spend the next eleven years of his life at the Cavendish. His experiences there and the people he worked with profoundly affected his future career. For these reasons, it is worth reviewing the history of this institution in some detail.

Founded in 1871, the laboratory was established in an attempt to recapture the Cambridge tradition of excellence in the experimental sciences that had largely disappeared since the time of Sir Isaac Newton. Even Newton had regarded experimental research as an activity to be pursued by the scientist alone rather than in a laboratory setting.[9] Money donated in 1871 by Henry Cavendish, duke of Devonshire, established both a laboratory and a professorship of experimental physics. The first appointment to the Cavendish chair was the Scottish physicist James Clerk Maxwell.

Maxwell remained in the chair until his death in 1879. Under his guidance, the use of experiments to demonstrate physical facts and laws became part of the teaching method. Maxwell's successor, Lord Rayleigh, continued these traditions but retired from the chair after only five years. Although the number of research workers increased during Rayleigh's tenure, their numbers were still small and few prominent physicists emerged.[10]

Choosing Lord Rayleigh's replacement proved to be difficult. The choice, when made, was a surprising one: Joseph John Thomson was only twenty-seven when elected to the directorship of the laboratory. His appointment proved crucial for the future of the Cavendish, however, for it was under Thomson that it grew to become a leading laboratory in experimental physics.[11]

Thomson brought important new qualities to the Cavendish. His middle-class background meant he was closer socially to many of the incoming research students; his experience in engineering lent an air of practicality and industry to the laboratory. During his first decade in the director's chair the volume of work accomplished and the number of research students doubled. The laboratory obtained a more solid foundation and was no longer solely dependent upon the individual genius of scientists like Rayleigh and Maxwell. Thomson was also responsible for forming the Cavendish Physical Society, which gave students a chance to discuss research and mingle socially. Equally important were the new regulations that, after 1895, allowed research students from other universities to come to Cambridge. Ernest Rutherford from New Zealand was the first such student to arrive at the Cavendish. In just under twenty-five years Rutherford would replace Thomson and become the powerful director of the Cavendish laboratory known to the young undergraduate Wilfrid Bennett Lewis.

Born in New Zealand in 1871, Ernest Rutherford excelled scholastically from an early age. Scholarships allowed him to pursue his education in New Zealand and, in 1895, to travel to England to work with J.J. Thomson. It was an exciting time to be at the Cavendish. Two years after Rutherford's arrival, Thomson made his famous discovery of the electron and measured the ratio of its charge to mass. For this work Thomson would later receive the Nobel prize in physics. By this time, Rutherford had advanced from his early work on the newly discovered X-rays to an examination of the element uranium. He would pursue this work in his new position as the Macdonald Professor of Physics at McGill University in Montreal, Canada.[12]

Rutherford remained in Montreal from 1898 until 1907. He devoted most of his time there to an examination of radioactivity – work he continued after his return to England in 1907. From then on he ceased solely to analyse radioactivity as a phenomenon, instead using it as a tool to probe the atom.[13] It was this approach that led to Rutherford's greatest triumph: his discovery, in his own words, of "what the atom looks like." Experiments conducted at Manchester in 1911 enabled Rutherford to present the idea of the nuclear atom. Rather than envisaging the atom as a mass of positive charge with electrons embedded throughout (Thomson's "plum pudding" model), Rutherford suggested that all the positive nuclear charge is concentrated at the centre of the atom with the electrons orbiting this "nucleus" in much the same way as the planets orbit the sun. As a simplified vision of the atom this description is still used today.

In 1919 J.J. Thomson resigned as Cavendish Professor. With Rutherford's reputation as a first-class experimental physicist firmly established from his work both at McGill and Manchester, it was not surprising that he was asked to succeed his old professor. Rutherford found the Cavendish in dire need of funds for staff, supplies, and laboratory space. In the hope that financing might be secured from the government, he wrote a report in which he outlined many of the achievements of the laboratory. Noting the contributions made by scientists to the war that had just ended, Rutherford argued the need for pure as well as applied research. The country, Rutherford held, would need physicists both for national work and for teaching physics in the schools. As David Wilson has noted, Rutherford would later be proved correct when, during the Second World War, physicists from the Cavendish were central in the development of radar.[14] But at the time, Rutherford's pleas for money and support fell on deaf ears. Funds for scientific research did not begin to arrive until the mid-1920s and were always limited. The resulting stringency had an adverse effect on the laboratory and is remembered by all those who worked at the Cavendish during those years.

When Wilfrid Bennett Lewis arrived at Cambridge in October 1927, he enrolled in a three-year honours degree course. During the first two years he studied physics, chemistry, and mathematics. At the end of this period the students were required to pass an exam, the Natural Sciences Tripos Part I. In his final year Lewis specialized in physics and sat Part II of the Natural Sciences Tripos. In both these exams he was awarded second-class honours.

At the Cavendish, lecture theatres and laboratories for the undergraduates were situated in the same buildings as the research activities. This meant that from the beginning, students were made to feel part of the laboratory and its work. A student who entered as an undergraduate four years before Lewis recalls that the Cavendish was then still very much Maxwell's creation. He remembers being "awed by a strong sense of tradition."[15] Unfortunately, the laboratory had remained "Maxwell's creation" largely for lack of funds. The tight budget affected all aspects of the laboratory including the undergraduate lecture courses. In the practical-physics class the experiments were always homemade. Freshman researchers had to make their own apparatus using hand- and foot-operated tools and "bits of metal and wood that had been used and reused by generations of research students."[16] This has become known as the "string and sealing wax" tradition of which many are proud, though it should also be recognized that this rugged approach must have held up valuable research.

In the final year of the undergraduate degree, students who wished to carry on to graduate work were asked to write down their research interests. Lewis recalled that he asked to do research in physics, but "in anything other than radioactivity." This resolve later vanished when Rutherford summoned Lewis to his office and said, "I am told you understand about these wireless valves. We are just beginning to use them in our alpha-ray work. If you get through your exams all right I would like you to join our group."[17]

Lewis accepted Rutherford's offer when he graduated with an honours B.A. from Cambridge in 1930. He continued at the Cavendish as a graduate student funded by a Department of Scientific and Industrial Research (DSIR) junior research studentship. All new graduate students began their tenure at the Cavendish by spending some time in "the Nursery" under the supervision of James Chadwick. It was here that new students learned the basics of experimental technique. Topics included experimental work with radioactive sources, glass-blowing techniques, and high-vacuum work.[18] Everyone also had to do some scintillation counting, "to see if you could do it," in the words of one Cavendish graduate.[19] The scintillations were flashes that occurred when alpha particles (heavy, positively charged particles) struck a zinc sulphide screen. Researchers counted the number of flashes that appeared in a given time period and used the results in their calculations. Scintillation counting was crucial to many experiments performed at the Cavendish throughout the 1920s and early 1930s. However, this method had distinct disadvantages. It depended heavily on the self-discipline of the observer, who could only be relied upon to count accurately for a short period of time. Often the procedure was for one observer to count for only a minute before being relieved by another. This would be repeated over and over during the course of an experiment.[20] John Cockcroft, a Cavendish physicist who would later figure prominently in Lewis's life, recalled practising with a partner. As he wrote to his wife, "two of us look at the screen together and when we see a flash, press a key and mark a tape machine outside. The game is to find out how many each miss and it appears I miss about one in ten and Boyce about 16 in 100."[21] Obviously eye fatigue and human error on the part of observers could lead to inaccurate results. Surprisingly, these screens remained the only method of particle counting at the Cavendish throughout the 1920s. This situation was beginning to change when Lewis entered the graduate program.

The disadvantage of the scintillation counting method was the degree of inaccuracy that arose from its reliance upon human observation. Rutherford, however, liked being able to "see" the alpha particles impinge upon the screen. But when results from the Cavendish were challenged by a team from Vienna, with both sides using scintillation counting, it was recognized that a more reliable method had to be found. Rutherford turned to a young Welshman in his group, C.E. Wynn-Williams, to develop an electrical method of counting.[22]

Interestingly, Rutherford himself had developed a method of electrical counting many years before in Manchester. In 1908, he and a colleague, Hans Geiger, had built a device that amplified the ionization effect of individual alpha particles so that their arrival would cause an electrometer to swing through fifty divisions, an effect that could be seen and counted. However, they found that scintillation counting, when compared to this method, was every bit as accurate. Having determined that there were now two independent methods of counting alpha particles, they abandoned the electrical counting method for the next two decades.[23]

Electrical counting, when it reappeared in the late 1920s, was more sophisticated in form. In 1928, Wynn-Williams tested the method of the physicist Greinacher, who, a few years earlier, had "succeeded in detecting individual alpha-particles and protons by amplifying linearly the ionisation currents from a few millimetres of their tracks in air." Satisfied with the results, they improved their counter and linked it to an oscillograph so that "not only could alpha-particles be counted at high speed even against a strong disturbing background of gamma-radiation, but it was also possible to distinguish, by their different sized deflections, the alpha-particles of different energy groups."[24] Rutherford, working with Wynn-Williams and F.A.B. Ward and later joined by Lewis, used this equipment to examine long- and short-range alpha-particle groups emitted from various substances. But problems still remained. The major drawback of the new method was that the results were recorded photographically and were therefore not immediately available; as a result, samples with a short active life were useless by the time the outcome of the experiment was known. This difficulty was overcome with the introduction of the thyratron to the Cavendish. When triggered by a voltage pulse from the amplifier output, a thyratron released an arc current strong enough to operate a mechanical counting meter. Wynn-Williams recalled that Rutherford was fond of using the single-thyratron counting device at Royal Institution lectures: "The changing dial figures, the sharp click of the mechanism and the bright flash of the arc, all helped to convey to the audience that alpha-particles really were being counted."[25]

The thyratron counting method replaced photographic recording; in the words of Mark Oliphant, a fellow Cavendish research student, it "revolutionized the rate at which statistically significant results in nuclear physics were obtained in the laboratory."[26] The new counting machines were introduced in the early 1930s and continually improved upon. They rapidly became a vital component in many of the experiments underway and were central to Lewis's future research.

In the history of physics, the year 1932 has been correctly labelled an *annus mirabilis*. Following upon Chadwick's discovery of the neutron, Cockcroft and Walton demonstrated for the first time the possibility of artificial disintegration of elements, and later in the year Blackett and Occhialini discovered the positron. As Rutherford commented at the time, "It never rains but it pours."[27]

And pour it did. The Cavendish laboratory in the early 1930s was an exciting place to practise physics. Famous discoveries were made and eminent physicists visited regularly to lecture and discuss the new results. Visitors bringing news of work at other university laboratories were often asked to speak at a meeting of the Kapitza club. Founded in 1922 by the visiting Russian physicist Peter Kapitza, the club was planned as a forum where the latest work in experimental physics could be freely discussed. Kapitza admired the research tradition existing at Cambridge but found that English research workers were often more reticent in putting forward new suggestions and ideas than their European counterparts. At club meetings, younger members of the laboratory and students were encouraged to participate. Soon the Kapitza club had gained a reputation as a leading physics seminar.[28]

It was into this stimulating environment that Lewis entered as a graduate student in 1930. Already familiar with the Cavendish and its inhabitants, Lewis did not hesitate to become involved in laboratory activities. In his first year as a doctoral student he was a visitor at Kapitza club meetings and by 1932–33 was listed as a member. His participation in the club lasted until he left the Cavendish in 1939.[29] At the time Lewis joined Rutherford's research team, it was made up of C.E. Wynn-Williams and F.A.B. Ward. Lewis recalled, many years later, that he had been asked to join this group because of his knowledge of electronic-circuit engineering; one of his first tasks was to help rebuild an amplifier. Lewis then participated in the use of their range-measurement technique to study the alpha-particle groups emitted from various radioactive nuclides. Altogether the group published six papers in the *Proceedings of the Royal Society* between 1931 and 1934, contributing important evidence for the quantum theory of radioactive decay then being developed by theoretical physicists.[30] Lewis's thesis, for which he was awarded a master's and a doctoral degree in 1934, grew out of this work. The thesis focused on the accurate analysis of alpha-particle groups and drew on the experimental results and techniques developed while working with Rutherford's group.

After four years at the Cavendish Lewis had an established reputation as the resident electronics wizard, a title he shared with C.E. Wynn-Williams. The fascination with electronics that extended back to his public-school days had, it seemed, served him well. His contemporaries at the Cavendish remember his electronics skills well. A.G. Ward, who overlapped with Lewis at the Cavendish and worked with him during the war and later at Chalk River, recalls that he was known as the electronics expert.[31] Another Cavendish contemporary, L.G. Cook, remembers that whenever anything went wrong with the apparatus, it was Lewis who was called in to find the problem.[32] As Cockcroft later put it, Lewis could "smell which valve or resistance in a set [was] giving trouble."[33]

Lewis's ability to track down and repair electronic breakdowns in the laboratory apparatus was a critical asset. A great deal of the work of an experimental physicist, then and now, lay in building the necessary equipment and ensuring that it functioned reliably. As noted earlier, the equipment employed at the Cavendish in the early 1930s was increasingly electronic in nature. Much of it was still in the early stages of development and was therefore particularly prone to problems and breakdowns. Because of the novelty of these devices, little use was made of manufacturing companies to obtain the instrumentation. This meant that Cavendish scientists were responsible for both the design of the apparatus and its construction. This type of practical, hands-on experience would be useful to the many Cavendish researchers, including Lewis, who, during the war, played vital roles in the development of radar.

In 1934, after he had received his doctorate, Lewis was awarded a research fellowship from Gonville and Caius College, which allowed him to stay at the Cavendish to conduct further investigations. He was later made a university demonstrator in physics and in 1937 a university lecturer. His research at this time included an attempt, made in collaboration with B.V. Bowden, to detect gamma radiation excited by the impact of alpha particles on heavy elements. Although the attempt was unsuccessful, better results came from an experiment performed by Lewis and John Cockcroft, in which they disintegrated certain light elements by bombarding them with protons and deuterons accelerated to high voltages. The results of both of these experiments were published, but increasingly Lewis seemed to be concentrating on work that was less experimental physics than engineering. When, in 1936, the Cavendish decided to add a new high-voltage laboratory, Lewis developed the electrical-counting technique, which introduced high-speed electron valve counters, simplified amplifiers, and counters for Geiger-Müller tubes. In the same year Cockcroft finally secured Rutherford's agreement for the construction of a cyclotron. The radio-frequency systems needed for this complicated machine were also largely Lewis's responsibility.[34]

Lewis's fascination with electronics led him to spend much of his spare time outside of the laboratory studying circuits and valves. His interest in wireless technology continued unabated; between 1931 and 1937 he published four articles in the magazine *Experimental Wireless and Wireless Engineer*. Technical in nature, these articles looked at certain problems in wireless technology and, through diagrams and mathematical equations, demonstrated methods of improvement. It seems clear from the complexity and length of some of these pieces that a great deal of time and thought was devoted to them.[35]

It is likely that Lewis's interest in wireless also led him to become involved in the Cambridge University Officers Training Corps (OTC). This was a voluntary organization that aimed at training men as part of the Territorial Army. Most public schools and universities had an attached OTC and Lewis had been a member of the Haileybury Corps. In 1931 he was commissioned as a second lieutenant in the Territorial Army. Soon he was working in the research section of the Signals Unit of the Cambridge University OTC. His main project from 1931 to 1938 (during which time he was promoted from second lieutenant to captain in command of the signals unit) was to develop duplex telephony on ultra-short waves – effectively a primitive walkie-talkie. The main problem lay in devising an instrument that could act as both receiver and transmitter; in other words, one in which conversation could pass in both directions at the same time.[36] Experiments were carried out on a rifle-range that lay on the outskirts of Cambridge. One of the huts on the range housed Lewis's laboratory, where he did much of his design work. J.C. Kendrew and S.W.H.W. Falloon, undergraduates at Cambridge, both joined the Signals section of the OTC and quickly became involved in Lewis's experiments. Lewis needed someone to operate a set at a considerable distance to see if the voice was being transmitted clearly. According to Kendrew, the climax to these experiments came during the annual athletics meeting when Lewis managed to convince the higher authorities that they would provide the radio communication from the middle of the stadium to the telegraph. Apparently this was a success and, as Kendrew commented, "quite entertaining."[37]

The time Lewis spent at the Cavendish laboratory in Cambridge was fruitful and productive in many ways. Both as an undergraduate and a graduate student he worked with many of the great men in nuclear physics, Ernest Rutherford, James Chadwick, and John Cockcroft, to name a few. Working at the Cavendish during one of its most exciting periods meant that he was able to absorb the atmosphere of an academic scientific laboratory at its best. He made connections with physicists, chemists, and other scientists that would be useful later in his career.

Lewis was also fortunate in his timing. He arrived as a graduate student at a crucial time when the older and simpler zinc sulphide screen counting devices were being replaced by a more advanced and more accurate electronic variety. He quickly established himself as an electronics expert and played an important role in developing the electronics in new experimental equipment. In 1942, many of the important innovations introduced by Lewis and others were published in his book *Electrical Counting*.[38] Finally, Lewis pursued in his spare time the study of wireless technology, both alone and through the Signals section of the OTC. In the end, it was the time Lewis spent upon electronic devices and wireless technology, rather than the experiments on radioactivity, that best prepared him for the work he would take up to serve his country during the Second World War.

CHAPTER TWO

Radar Days

In 1938, Wilfrid Bennett Lewis turned thirty years old. For someone so young he had already achieved a great deal. His doctorate in physics, completed four years earlier, had been awarded by the most prestigious university for experimental physics in Britain, and perhaps the world. The Cavendish had fostered a succession of famous physicists, many of whom were there when Lewis was. Lewis was also fortunate in receiving grants and lectureships at Cambridge so that, although his long-term future career remained uncertain, he was able to live comfortably.

Friends and colleagues describe Lewis at this time as a shy man, sometimes appearing aloof. Not very tall, he was heavily built. Fellow workers recall his enormous appetite for food, which often resulted in his finishing meals long before others at the table. Although generally well liked by his colleagues, it would seem that Lewis enjoyed being alone, working hard day and night.[1]

Perhaps Lewis's most important attribute was his ability to devote himself completely to the task at hand. This capacity to concentrate his efforts towards a single goal, combined with his extensive knowledge of electronics and his connections in the university community, made him well suited to the task presented to him on the eve of World War II.

The Second World War has been called "the physicists' war" because of the enormous extent to which physicists participated in developing offensive and defensive weapons for the three services of the British armed forces. The label is unfair in that it neglects the many chemists, biologists, mathematicians, and engineers who were also involved, but the main idea is clear: during World War II scientific expertise and ideas were linked to military needs to an unprecedented extent.

That this should occur was by no means obvious. Before World War I, little had been done to use scientific knowledge to improve military weapons. But as the conflict that was to have been "over by Christmas" dragged on and the national war effort intensified, civilian scientists were drawn into military work on aircraft, submarine warfare, tanks, poison gas, and other subjects. In 1916, in a formal attempt to mobilize science for war, the British government established the Department of Scientific and Industrial Research with the mandate to make the best use of the country's resources for the war effort.[2]

DSIR survived the war and, in the early postwar years, coordinating boards were set up within the department in an attempt to link the work of the research establishments with the armed services. But in the 1920s expenditure on the services was a political liability; in 1927 all of these boards except for the Radio Research Board were disbanded.[3] By 1930, very little contact between civilian scientists and the military remained.

A large part of the difficulty lay in attracting scientists to work on military problems. Many hoped that there would not be a second war and so were not interested in becoming involved in the preparation for one. Furthermore, it is doubtful whether the funds would have been available to pay a large number of scientists to prepare for another conflict at a time when the majority of the population fervently hoped that war could be avoided. Finally, the secrecy necessary for armaments research was anathema to most scientists, who enjoyed the spirit of international cooperation that existed in the scientific world.[4]

But as the decade progressed and the likelihood of war grew, Britain's potential vulnerability to air attack became a cause for concern. In a speech made in the House of Commons in November 1932, Stanley Baldwin stated his belief that "the bomber will always get through."[5] The summer air exercises of 1934 seemed to confirm this aphorism: a simulated attack completely destroyed both the Air Ministry and the Houses of Parliament. In the face of growing German air power it became clear that, unless some means of improving Britain's air defences were found, chances of winning the next war were very slight.[6]

It fell to A.P. Rowe, at this time "the only member of the Headquarters' staff of the Director of Scientific Research, Air Ministry, employed wholly on armament problems," to act on these concerns for his country's future.[7] In June 1934 Rowe gathered together all the Ministry files available on air defence – a total of fifty-three. He quickly realized that although some thought had been given to the issue, no scientific approach had been made. Rowe summarized the situation in a memo to the director of Scientific Research, suggesting that unless scientists were able to devise a new method of defence against air attack, Britain would lose any war starting in the next ten years.

In response to Rowe's memo the director, H.E. Wimperis, suggested that a committee under the direction of Henry Tizard be set up to consider the problem. Tizard was a good choice. In the early post-war years, it was Tizard who had been brought into DSIR in an attempt to coordinate the work of the research establishments with the needs of the services. He had a reputation for working well with both scientists and members of the armed forces. The other members of the committee were A.V. Hill, a physiologist who had worked during the First World War as director of the Anti-Aircraft Experimental Section, Munitions Invention Department; P.M.S. Blackett, a naval officer before and during the war who then switched to a scientific career at the Cavendish in Cambridge; H.E. Wimperis; and A.P. Rowe as secretary.

The terms of reference of the Tizard committee, as it soon came to be called, were "to consider how far recent advances in scientific and technical knowledge can be used to strengthen the present methods of defence against hostile aircraft."[8] Before beginning such an investigation, Wimperis wanted finally to establish whether any scientific validity could be attached to the idea of a "death ray." Over the years the Air Ministry had been plagued by inventors claiming to have developed a ray capable of killing a sheep, for example. With more time and more money, they argued, they could develop a ray that could destroy an aircraft. As fantastic as this idea seemed, Wimperis felt it necessary to ensure that no such weapon was in fact physically viable.[9] He telephoned Robert Watson-Watt, at the time the superintendent of the Radio Research Laboratory of the National Physical Laboratory, and asked him if a form of damaging radiation was possible as a defence against enemy aircraft.[10] Watson-Watt asked his assistant, A.F. Wilkins, to perform the necessary calculations. Wilkins quickly showed that a death ray was not physically possible. Watson-Watt then asked Wilkins "to find out what power would be required to produce a detectable signal from an aircraft at such and such a range."[11] This calculation was more promising. In a memo sent to Wimperis, Watson-Watt reported the possibility of detecting an aircraft by reflecting radio waves off it. This reply arrived shortly before the first meeting of the new Tizard committee.

The fact that radio waves could be reflected off objects had been recognized for a long time by scientists in many different countries.

In an address given in 1922, Marconi had stated that a device to detect ships by bouncing radio waves off their hulls could be designed. In England, E.V. Appleton had been measuring the height of the reflecting ionosphere by bouncing radio transmissions off it. But the birth of radar as a practical technique can only be said to have occurred when Watson-Watt linked the scientific possibilities with the operational needs of the Air Ministry.[12]

After this, events moved quickly. Wimperis obtained ten thousand pounds for investigating the new method of detection and by the end of February 1935, a demonstration of the new technique to the Tizard committee had been arranged. On 26 February 1935, Rowe, representing the committee, watched as an observable reflection was received from an aircraft as it flew through a local radio beam at Daventry.[13] This experiment confirmed the premise upon which subsequent radar systems would be built: electromagnetic radiation, in the form of radio waves, could be bounced off aircraft and the waves reflected back were strong enough to be received. Since radio waves travel at a constant speed, the range of the aircraft could be easily calculated.

After the success of the demonstration at Daventry, the next step in the evolution of radar was to set up a research station to develop the technique. In May 1935 a small group of scientists began work at Orfordness, an isolated site on the east coast of England five miles south of Aldeburgh. Their early objectives were straightforward: to determine the position and height of the incoming aircraft, their numerical strength, and, perhaps most importantly, whether they were friend or foe.[14]

The first problem to be tackled was increasing the range of detection. This was done by raising the heights of the towers used to transmit and receive signals. By July 1935 signals had been received from aircraft flying forty miles away. In a memo to the Tizard committee, Watson-Watt suggested that a chain of towers be erected along the coast to warn of incoming aircraft. The Air Council accepted this idea and the construction of twenty chain-home stations began. In the words of the official historians, this was "a strategical commitment of the first order."[15]

Before these stations could be built, further advances were made so that angle of bearing and elevation of the aircraft could be measured from a single station. Within a few months the scientific modifications necessary for the construction of chain-home stations were effected at the research station at Orfordness. Located on a long spit of land, the Orfordness research station was small and primitive. It was clear that for further development of radar to take place a more substantial experimental station would have to be found. A search for suitable locations turned up Bawdsey Manor, twenty miles down the coast from Orfordness. Sitting on a seventy-foot promontory rising out of the surrounding flat land, the manor and its grounds were well suited for a scientific research establishment. With its varied architectural styles and its "spacious lawns, peach trees, [and] bougainvillaea bushes," the manor provided lovely surroundings for the intense scientific work taking place within.[16]

The move from Orfordness to Bawdsey was completed by May 1936. The establishment, now known as Bawdsey Research Station, would remain there until the outbreak of war. Much of the work done during this period was connected with the construction of the chain-home stations. In his third memorandum to the Tizard committee, Watson-Watt had suggested that the best way to defend the coasts of England would be to "floodlight" them with radio beams. The wavelength of the beams used at this time was very long (around twenty-six metres), which meant that low-flying aircraft might not be detected. But Watson-Watt argued that this system should be put in place while shorter wavelengths of sufficient power were developed.[17] As will be seen, a large part of radar development during the war was the drive towards shorter and shorter wavelengths.

Shorter wavelengths were also essential for any kind of airborne radar system. With the chain-home system under construction, the problem of defending the country against daytime bombers seemed to have been solved. Incoming bombers would appear on the radar screens and fighters could be guided from the ground to within visual range. The potential problem of German night bombing remained, however. To meet this challenge, each airplane would have to be individually equipped and shorter wavelengths would be necessary for the fighter to get close enough for visual contact at night. The miniaturization required proved to be enormously difficult. Radar equipment in the chain-home stations filled rooms and weighed tons. E.G. Bowen, a young physicist who had been at the Orfordness research station, began to examine this problem shortly after the move to Bawdsey. By August 1937, Bowen and his small team were receiving echoes off boats two miles away using an airborne system.[18]

Bawdsey Research Station was still, at this time, a modest establishment. It was difficult to convince many scientists to do secret, war-related research during peacetime. But growth did occur. In May 1938, Watson-Watt moved from his position as superintendent of Bawdsey to become the director of Communication Development at the Air Ministry. He was replaced by A.P. Rowe, secretary to the Tizard committee. A dedicated civil servant with limited understanding of the technical details of the work, Rowe would remain superintendent until the end of the war. At the time he became director, the population of the research station was 150 people, including the cleaners.[19] This number would grow, however, as the international situation became more unstable.

On the eve of the war, the state of radar presented a mixed picture. On the one hand, work on the chain-home system had proceeded expeditiously: in September 1939 the stations were operating around the clock, ready to provide timely warning in the event of daytime attack. Night-time defence was another matter, however. As noted above, air-interception work remained in the experimental stages, with the result that no suitable form of night defence was in place at the outbreak of war. Moreover, this work had not yet been extended to attacks against submarines, a surprising deficiency given the crucial role air-to-surface vessel radar would later play in battles on the sea. This shortcoming was rooted in the limited number of scientists available to work on these problems, which in turn reflected the difficulty of recruiting scientists into defence work. By the spring of 1938, it was clear that more scientists would have to be recruited. The obvious place to look for these men was in the universities.[20]

Bawdsey Research Station and the Cavendish laboratory at Cambridge University were located only a short distance from one another, but thus far there had been no communication between the two establishments. This was due in part to the secret nature of the work at Bawdsey. But it was compounded by the absence of any mechanism through which contact could be made. By the spring of 1938, Henry Tizard realized that many more physicists and engineers would be needed for further research and development and to help look after the radar stations in the chain home. He contacted John Cockcroft at the Cavendish and, over lunch at the Athenaeum, told him about the new, secret radio technique for detection of enemy aircraft. Cockcroft, as he later recalled, was told that "these devices would be troublesome and would require a team of nurses – would we – the Cavendish – undertake to come in and act as nursemaids, if and when war broke out."[21]

After this early contact, little action appears to have been taken over the summer. In fact, on 13 September 1938, only two weeks before the Munich conference, Tizard wrote to the new Cavendish professor W.L. Bragg (Rutherford died in 1937) to tell him that while the final decision about introducing scientists to radar secrets had been put off, lists of eligible scientists should perhaps be prepared.[22]

Although the agreement signed at Munich at the end of September defused the immediate threat of war, the Czech crisis spurred the scientific community into action. Shortly after Munich, Cockcroft and

R.H. Fowler, another Cavendish physicist, drove to Bawdsey to be briefed on recent developments in both ground and airborne radar. The desire for secrecy, however, slowed the process of introducing a large number of scientists to the problems of radar. It was not until 1 February 1939 that a high-level meeting was held with Cavendish physicists to discuss organization for war.[23]

Although this meeting may have been Lewis's first introduction to radar problems, he had been mentioned many months earlier as an important scientist for radar work. With his expertise in electronics technology, his fascination since childhood with radio and its components, and his technical experiments carried out with the Cambridge OTC, Lewis was an obvious radar candidate. In October 1938 Cockcroft had told Tizard of Lewis's interest in short-wave radio work. Tizard instructed Cockcroft to tell Lewis of their interest in "high powers in short waves." Any ideas Lewis had about this would be welcome.[24] But concerns about secrecy remained; at the end of November 1938 Tizard still had to ask Watson-Watt if Cockcroft could speak of the anti-aircraft problems with his colleagues Lewis and P.I. Dee.

W.L. Bragg chaired the February 1939 meeting of Cavendish scientists that included W.B. Lewis, J.A. Ratcliffe, and P.I. Dee. At the meeting a list of scientists from a number of universities was compiled. These men were then divided into groups of seven or eight. Each group would be sent to a different station where, after one week of instruction by the permanent staff, they would run the station themselves. It was planned that this training program would be put into place between the end of August and the middle of October 1939.[25]

Lewis was a central figure in this early organization of scientists for war work. In the spring of 1939, he accompanied Cockcroft and two other Cavendish researchers on a visit to Bawdsey. In subsequent meetings and discussions, as the official historians note, "it was agreed that Dr Lewis should go to Bawdsey to take charge of a wide field of research."[26] From the beginning of July, Lewis worked at Bawdsey on a part-time basis with the intention that he would work there full time by October. He spent the summer acquainting himself with the details of radar. His rapid comprehension of the technical difficulties is evident in his early reports.[27] Other duties included recruiting more scientists into radar work – part of the training scheme that had been planned for the month of September. This plan was disrupted by the outbreak of war with Germany on 3 September 1939.

The war intruded immediately and directly upon the work of the radar scientists by forcing a change of location. It had long been recognized that radar stations were obvious targets for German bombs.

With their tall towers, it was expected that, once apprised of their purpose, the Germans would attempt to destroy them. Bawdsey, the centre of radar research in England, had many such towers. It had therefore been decided that on the outbreak of war, the station would move to Dundee, Scotland.

The location was far from satisfactory. A lack of space forced certain teams to relocate elsewhere, thus breaking up the research effort. Within a year a new location had been found on the south coast near Swanage, but this too proved to be short-lived. It was not until 1942 that the establishment settled in its final location of Malvern, Worcestershire. There were also changes in name. In Dundee the establishment was called the Air Ministry Research Establishment; this was later changed to the Ministry of Aircraft Production Research Establishment. But the final name was the Telecommunications Research Establishment, or TRE.

Despite two changes of location in less than nine months, much was accomplished in the development of radar during the "phoney war." Work continued on air-interception sets, which helped fighter planes home in on incoming bombers. The main problem, that of the minimum range of detection exceeding the maximum visual range, was solved during this period through collaboration with Electrical and Musical Industries (EMI). Scientists designed and constructed a Plan Position Indicator, a screen showing the position of aircraft on a grid. Air-to-surface vessel radar received further attention, a development that would prove crucial later in the war. TRE scientists also turned their attention to developing offensive radar systems. Up to this point, most of what had been developed was defensive radar, systems that would protect Britain against incoming bomber attacks. But in June 1940, shortly after the establishment moved to Swanage, Air Marshal Joubert visited TRE and outlined Bomber Command's pressing need for a device that would help bombers hit their targets. Developed by R.J. Dippy, the navigational system called "Gee" would be enormously helpful in directing RAF bombers to their targets.

With these larger development goals in mind, scientists at the establishment built and refined radar components. Lewis led a group responsible for "basic design work on transmitters, receivers, display systems, direction finding, anti-jamming and training apparatus."[28] Initially Lewis both directed research and remained actively involved in it himself. Most of his technical papers on radar appeared in this early period. Later he became responsible "mainly for fostering and planning research and development."[29] Because of his seniority at the Cavendish, Lewis entered the radar establishment at a high level, that of senior scientific officer, and he quickly advanced to the position of principal scientific officer in April 1940. A year later his promotion to assistant superintendent formally acknowledged his role as Rowe's deputy. When Rowe became chief superintendent in 1943, Lewis moved to the position of superintendent, where he remained until the end of the war. These rapid promotions were not appreciated by some scientists who had been involved in radar longer, but Chief Superintendent Rowe was adamant. Rowe apparently recognized Lewis's ability to grasp the details of the varied scientific and technical problems involved in radar development. While Rowe dealt with the political problems of the establishment, Lewis looked after scientific administration. Rowe later acknowledged that he had "risked offending my old colleagues by appointing Lewis as my deputy although he had been with us but a few months."[30]

Another factor in Lewis's rise in the establishment was its rapid expansion to meet the growing radar requirements of the deteriorating military situation. After the fall of France in June 1940, the German *Luftwaffe* relocated to French airfields, thus posing an even greater threat to Britain and underscoring the need for more radar scientists. There had only been four hundred people to move from Dundee to Swanage but by the end of the war the population of the establishment was close to three thousand. Scientists were divided into various groups, which in turn made up larger divisions. A senior scientist headed each division. Initially, Lewis was one of seven or eight division heads but he soon emerged as the senior scientist.[31] As a division head he was still actively involved in research; he both guided the research efforts of those in his groups and did work himself. It is difficult to pinpoint exact scientific contributions made by Lewis as he was involved in many different joint projects. In general, however, as the size of the establishment grew, his role became more one of scientific and technical direction.

Although the groups that he guided worked on a wide variety of problems, Lewis's main interest was in the electronics of transmitters and receivers. Despite his increasingly administrative role, he continued to write technical reports and summaries of work for Rowe. Committee work began increasingly to occupy his time. Many of the committees were highly technical and some involved interaction with industry. This was essential if the newly developed radar components were to be mass produced in time for their effective use in the war.

As the war progressed and greater liaison was needed between the developers and users of radar, Lewis began to serve on interservice committees that brought together the scientists, representatives from industry, and air-force personnel. These renowned meetings, known as "Sunday Soviets," earned their name from the fact that rank and seniority were ignored in the proceedings. First suggested by Rowe, the Sunday Soviets played a vital role in the rapid development and deployment of many radar systems. The tradition started while the establishment was located at Swanage but continued after the move to Malvern. Their importance in bringing together the developers and users of the radar systems cannot be overstated. The significance of these meetings was linked to the changing nature of TRE. Although originally set up as a radar research establishment, the exigencies of war soon meant that less basic research and more developmental and applications work was performed at TRE. This was not surprising. Once certain basic problems in radar research had been solved, little remained but to further refine these developments and adapt them to different radar systems. Furthermore, once developed, the radar systems had to be made comprehensible to the service personnel who would be using them in battle. In wartime, the luxury of performing basic research had to be curtailed.

The majority of the staff at TRE were research scientists who had to make the adjustment to development and applications. Many found this difficult. Lewis, however, seemed to be ideally suited to his new position. His time spent at the Cavendish had accustomed him to the practice of the senior professor walking around, visiting the various research projects in progress. Lewis adopted this technique and was thus able to involve himself in many aspects of work going on within the different divisions.[32] He had little time for the line organization, instead preferring to go directly to the man who could provide him with the necessary information. One colleague noted early on that Lewis was to be the "controlling and unifying brain" of the establishment. But the same colleague doubted that Lewis could maintain "dictatorial control over every other subject" from "techniques to the provision of bicycles and tea."[33] Clearly Lewis had problems in delegating responsibility to those beneath him. In his position as deputy superintendent of the largest research establishment in Britain during the war, this was not an insignificant drawback.

Lewis's Cavendish background proved enormously helpful in alleviating the difficult problem of recruiting scientific personnel. During the first few months of the war, the official Air Ministry procedure for appointing people was not yet well organized. This meant that some scientists who had spent time at Bawdsey and on chain-home stations were without instructions when the war began. Lewis became a useful contact for many scientists who wished to participate in war work but who did not know how to get involved. J.C. Kendrew, for example, had been in Lewis's OTC section at Cambridge. As he later recalled, "When the war came I was a chemist and nobody seemed to want me to do anything useful. Having heard that Lewis was involved in something very secret I wrote to him and he recruited me to TRE in January 1940."[34]

By virtue both of his scientific abilities and his time at the Cavendish, Lewis had the respect of the incoming university scientists. This would prove important in helping these men to adjust from work in universities to the more regulated atmosphere of a government laboratory. Lewis acted as a link between these newly recruited academics and the civil-service administrators above him.

Although these administrative aspects of Lewis's role were important, it was his technical expertise that was crucial. His scientific and technical qualifications were, of course, excellent. But perhaps more important, he was completely committed to the development of radar. In his memoir of the period, Rowe noted that for someone to advance to a high position in TRE, they needed to have a "fervent belief" that only through the full development of radar could the war be won.[35] W.B. Lewis was such a person. He was committed to the search for truth and felt strongly that he must use his talents to the utmost to help meet the national crisis then at hand.[36]

Fierce dedication had its drawbacks, however. For example, one area of work at TRE, called Radio Countermeasures, dealt with methods of jamming enemy radar to prevent German bombers from hitting their targets. The branch also explored methods of avoiding jamming by the Germans. Although it was clearly essential to the larger war effort, this research ran counter to what Lewis considered to be the first priority, the further development of radar itself. Robert Cockburn, the head of the radar-countermeasures section, recalls the difficulty of convincing Lewis that more staff was needed in this group. Radio Countermeasures felt starved of both staff and resources. Lewis apparently disliked committing precious resources to work designed to thwart the positive achievements of radar.[37]

Most of the radar equipment developed during the Second World War evolved out of basic discoveries made in the prewar and early war years. A major breakthrough in radar development occurred at Birmingham University in the fall of 1939. It was, in the words of the official historians, "one of the most striking and dramatic leaps forward in the whole scientific history of the war."[38] In the early days of radar, wavelengths of metres were used. But it was soon recognized that shorter wavelengths would allow better accuracy in sighting planes and ships. The problem lay in generating the power required for these wavelengths to be effective. In the fall of 1939, a development by two physicists at the University of Birmingham brought about what has been called "the centimetric revolution."[39] The development of the resonant-cavity magnetron meant that radar could be propagated with the power of kilowatts rather than watts and at wavelengths of ten centimetres or less. This was an enormous advance. TRE quickly moved to develop a program of centimetric radar.

Vitally important for the further development of more sophisticated radar equipment, the magnetron also became the centrepiece of a scientific mission to the United States in September 1940. The despatch of this "British Technical and Scientific Mission" (known as the "Tizard mission" after its leader, Henry Tizard) constituted an important development in Anglo-US relations. On the advice of Tizard, the British government had decided to divulge unilaterally the most important developments in radar in the hope of securing American (and, incidentally, Canadian) support and assistance in the increasingly demanding task of radar development and production.[40]

Lewis did not participate in the Tizard mission but prepared a memorandum comparing the positions of the two countries in radar development. Radar research in the United States, while ongoing, lacked the intensity of the British effort. Although certain systems were more advanced, in most respects the Americans trailed behind the British. The Americans were superior in making crystals and waveguides, but British work on receivers, transmitters, and time-base presentation was much further advanced.[41] The British disclosure of the magnetron, which increased the power available to the Americans a thousandfold, came as a welcome surprise. The Tizard mission led directly to the formation of the Radiation Laboratory at the Massachusetts Institute of Technology, where further developments in radar took place and production accelerated. The interchange thus proved useful to scientists on both sides of the Atlantic.

Perhaps because of the success of the Tizard mission, the British sent a second mission to Washington in 1943. This time Lewis was one of the active participants. Robert Watson-Watt led Lewis, Cockcroft, and representatives from the services. As with the Tizard mission there was an interchange of scientific work done in the two countries. Discussions centred on the use of mobile radar sets for campaigns in Europe and the need to avoid duplication in radar aids needed for invasion and for the war in the Far East.[42] Lewis also suggested that an attempt be made to establish a common standard for frequency allocation and design of equipment to make radar sets built in either country interchangeable.[43] Little has been written about this second mission to the United States, an undertaking that left Lewis disappointed. He felt that not enough attention had been paid to the information the British had supplied.[44] The episode was characteristic of the evolving inequalities in Anglo-American wartime scientific cooperation. The brush-off Lewis sensed, and the reasons behind it, were similar to those already experienced by his British colleagues at work on the atomic bomb. As with the atomic bomb, so with radar: as American scientific research efforts outstripped those of their allies, the Americans became less forthcoming in exchanging information.

It was clear from these two scientific missions to the United States that the Americans recognized the importance of the further development of radar. They could not but have been impressed by the British wartime achievements. As an island nation threatened by a strong German airforce, Britain had been forced to develop a system of early warning. The chain-home stations had proved crucial in ensuring the country's survival during the Battle of Britain. Advances in radar continued as the war progressed; by 1945 radar was being used for both defensive and offensive purposes. The success of the new developments and the speed with which they were brought from the laboratory into operation was remarkable.

Several features of TRE helped to account for this success. Quality of staff was obviously a crucial component. The majority of scientists working on radar were culled from the universities, which meant that most combined a high standard of ability with the theoretical outlook of the physicist rather than the engineer.[45] Many of the physicists came from the Cavendish laboratory, whose "string and sealing wax" tradition was well suited to the development of radar. The pressing needs of the different services in the midst of war meant that no time was available for refinements. Apparatus had to be put together with the materials available. Nothing could be wasted. These habits, learned at the Cavendish under Rutherford's tight rein, were put to effective use by Cavendish alumni at TRE.

But other research methods had to be modified. University researchers tended to focus on "basic" research, a task that Postan, Hay, and Scott describe as attempting "to find out how, in particular respects, nature works."[46] Practical applications can often be found for such work, but not always. If nothing but applied research is performed, however, there is a risk that new and important innovations might be missed. But in the conditions of national emergency under which TRE scientists worked, basic research with no practical application in sight might have been a waste of valuable men and resources. Clearly a middle road had to be found between these two extremes.

At TRE, this dilemma was dealt with by closely integrating the different stages from basic research to development and, finally, to applications. Thus the scientists would in some instances have to perform the basic research that would lead to an idea for a new radar device. They would then develop a prototype, ensure that it worked in an operational sense, and then collaborate with industry to have it mass produced. Finally, the servicemen would be introduced to the new device and taught how to use it. Much of this work was far removed from that to which most research scientists were accustomed. But again, scientists from the Cavendish (and likely from other universities as well) were used to constructing experimental apparatus and so were well equipped to deal with the first two stages. As the official history suggests, the TRE scientists' high capability in this regard must be considered as a factor in the success of the radar effort.[47]

A further dimension of the scientists' radar work was anticipating the tactical and strategical requirements of the services in devising new equipment. This led to a rise in the status of scientists; having proved their importance to the war effort, they were in a position to demand and receive greater attention. Under the impact of radar, the fledgling relationship between science and government that had existed before the war was transformed into a vital bond that would remain in place after 1945.

The establishment itself, TRE, was also key in the successful development of radar. Official historians Postan, Hay, and Scott have argued that certain features of the radar station at Bawdsey were vital to radar development and remained constant during the many moves of the establishment. Bawdsey Manor itself was an isolated yet self-contained world. With its large grounds and numerous rooms for laboratories, research and experiments could be performed without leaving the estate. As noted above, the scientists brought to work there hailed from a university environment, which added to their effectiveness. The novelty of the work and its importance to the operations of the RAF led to an enormous outpouring of new ideas. Although the physical surroundings of the establishment changed during the war, the spirit of Bawdsey was, in large part, maintained.[48]

W.B. Lewis was strongly influenced by the time he spent working at TRE. The atmosphere there was very different from that of the Cavendish laboratory. As discussed earlier, the work on radar was much more goal-oriented than that done in a university laboratory. But temperamentally, Lewis was well suited to the single-minded devotion to a cause necessary for this sort of work, and he would bring this same focused approach to his future scientific endeavours. Although Lewis only experienced the isolation of Bawdsey Manor for a few months before the establishment moved, much the same atmosphere could be found at Malvern, the final home of the radar establishment. With few distractions, the important job before him occupied Lewis to the exclusion of all else. The isolated work environment and complete submersion in research would characterize Lewis's postwar career in Canada as well.

It is clear that Lewis's experience in radar strongly influenced how he would manage future scientific establishments. At the Cavendish he had been one research worker among many, and although the work he did there was useful, he did not stand out. There the spark of individual genius distinguished people, while at TRE Lewis was able to call upon his electronics skills and develop his administrative abilities. He entered at a high level and quickly rose to the position of deputy chief superintendent. When at the end of the war Rowe announced his intention to leave, there was no doubt that Lewis would replace him.

Lewis's position at TRE as Rowe's deputy was an important one. Rowe was a professional civil servant who handled the political aspects of the establishment; Lewis provided the scientific back-up Rowe needed. Unlike Rowe, Lewis really understood electronics and when necessary could call on his scientific expertise to prove his point.[49]

This aspect of Lewis's role is best illustrated by the debate over the development of centimetric wavelengths. During the summer of 1940, centimetric radar development was still in its earliest stages. With the fall of France and the seemingly imminent invasion of England, Rowe experienced considerable pressure from his political and military superiors to concentrate work on immediately useful defensive projects. As Bernard Lovell, a colleague of Lewis's from TRE, has written:

> Indeed, it is understandable that the Service Chiefs and Government officials who visited TRE during the summer of 1940 at Worth Matravers, would find difficulty in understanding the relevance of the work of a group of newly recruited University scientists with this elementary 10 cm equipment and strange aerial systems. To Lewis, almost entirely, must go the credit of convincing Rowe that this influx of new talent must be concentrated on developments which could not possibly materialise in any time scale to deal with the immediate threats to the country.[50]

Ultimately, centimetric radar led to the development of a number of important radar devices, including H_2S, the blind bombing and navigational aid.

The end of the war in Europe brought about changes at TRE. Decisions had to be made about the type of work that should be carried on at the research establishment. In a memo written in May 1945,

Lewis noted that because the war in Japan was geographically remote, short-term work for immediate operational use had lost its urgency; the establishment should therefore concentrate more on basic research.[51] The end of the war in Japan, however, raised the question of whether TRE should continue to exist at all.

Lewis, among others, argued strongly that it should. His May 1945 memorandum proposed several peacetime applications for TRE's wartime accomplishments. Radar, for example, could be used for both civil aviation and the peacetime airforce. Advanced electronics systems developed during the war could be adapted for industry, in particular for atomic energy research. Furthermore, due to the pressures of war, many promising lines of research had not been pursued.[52]

In May 1945, the government agreed with Lewis's assessment of TRE's continuing importance. The minister of Aircraft Production, Sir Stafford Cripps, visited TRE to praise the scientists for their vigorous wartime efforts. The minister also announced that the government supported further work on radar. In particular, research would be performed to extend radar technology to shorter wavelengths and peacetime applications of radar would be explored.[53]

By October 1945, A.P. Rowe had resigned and Lewis had taken over as chief superintendent of TRE. Facing him was the difficult task of ensuring the survival of TRE and guiding it through the difficult reconstruction period. Since May, the previous strong government support had evaporated. A cabinet shuffle, a change of government, and a fundamental reorganization of wartime government departments had led to considerable confusion surrounding the future of the establishment. In an October 1945 memorandum, Lewis presented the case for continued strong government support of TRE. A major difficulty in sustaining support was that, because of the necessary cloak of secrecy surrounding radar during the war, public ideas about its origins were incorrect. The public, he claimed, believed that radar was discovered by a few government scientists and then developed by engineers and industry. This led to the conclusion that only a small number of scientists would be needed in the future. Such a view, in Lewis's opinion, was completely wrong. He recognized the need to widen the scope of the establishment's activities, in particular by moving away from military work, but argued that the government should not "disband or plunder the existing Establishment."[54]

Even without government "plunder," the establishment's uncertain future meant that retaining scientists was a serious problem. The majority of scientists had left jobs in universities to aid the scientific war effort and were anxious to return, especially if TRE had no future.

To let a highly trained and integrated group of scientists simply disband, Lewis argued, would be an enormous national loss. He recalled the integration of men with knowledge of strategic needs, practical needs, physics, and engineering and noted that such people would be required in peacetime England. Although scientists needed for fundamental research and teaching should return to those jobs, he suggested that 650 out of the wartime complement of 850 scientists be kept on.

Lewis's arguments were persuasive. Having gathered together an excellent team of scientists and engineers and built up contacts with industry, universities, and the services, it seemed foolish to let them disappear. Radar had many important peacetime uses and advanced electronics devices would be crucial in the years to come. But the new minister in charge of TRE, J.C. Wilmot, did not fully agree. Rather than supporting the idea that fundamental research on new radar systems should be continued at TRE, he countered that TRE "ought to have clear practical objectives." He added that "a Government department doing applied science without any real objective tends to drift."[55] This attitude did not bode well for the continued existence of TRE as a government research establishment.

Frustrated by the minister's refusal to follow his suggestions, Lewis appealed to Tizard for help. In a letter dated 16 December 1945, he expanded on the ideas put forward in his May memorandum. Especially interesting in light of his future career path were Lewis's suggestions that in the postwar years, "the objective is to produce aid for ventures which may appear rather long term economic propositions to British industry; examples are: civil aviation, where significant economic gain can only follow after establishing prestige through technical efficiency: 'Tube Alloys' [the code for the atomic-bomb developments] where economic gain follows on the development of the efficiency of processes of releasing and converting nuclear binding energy ..."[56] In March 1946 Lewis again wrote to Tizard to tell him that scientific staff at TRE had already been cut from a wartime peak of 880 to 660 and he had been ordered to make a further reduction of 250. But despite Lewis's plea for help, Tizard could only respond that he was no longer in a position to influence these decisions.

In yet another attempt to maintain the strength of TRE, Lewis wrote a memorandum in May 1946 arguing that the excellent contacts developed by the team at Malvern during the war with both industry and the universities could not easily be duplicated. Lewis noted that industry was still behind TRE in development of certain devices so that it would be mutually beneficial for the contacts already in place to be maintained. Numerous personal contacts existed between the universities and TRE. If these remained in place, the establishment could reap the benefit of close contact with the fertile minds of university scientists. He suggested that, rather than let TRE shrink in size, it should, through the absorption of parts of other establishments, grow in size and function.[57]

Although Lewis would not be around to witness it, this final suggestion predicted TRE's future. Eventually amalgamating with a number of other research establishments, it still exists today as the Royal Signals and Radar Establishment in Malvern, England.

In his final year at TRE, much of Lewis's time was spent pursuing administrative functions, as described above. But research continued at the establishment and Lewis remained involved in certain aspects of it. His most significant scientific contribution during this period was in the area of infra-red detection.

Interest in infra-red radiation (i.e., radiation having a wavelength just beyond the red end of the visible spectrum) had arisen during the war in connection with the schnorkel tube, developed late in the war for use by German submarines. By using this tube the submarines could remain submerged while recharging their batteries; this made radar detection virtually impossible. But it was possible to detect them by use of infra-red techniques.

Lewis's contribution to infra-red theory came after the war had ended. He tried to determine the limits of the infra-red techniques. In a paper he attempted to show the nature of its limitations but also proved that a thermal detector can be "a very sensitive receiver capable of responding to quickly changing scenes of natural thermal radiation." He later expanded this work into a paper on fluctuations in streams of radiation, subsequently published in the Proceedings of the Physical Society.[58]

Lewis remained as chief superintendent for only a year. During that period he guided "the establishment through the reconstruction period, leaving it as a composite establishment of three units." One unit continued with the development of radio and electronic aids for control of flying and air warfare. The second worked on electronic applications for atomic-energy research, while the third pursued basic electronics research in such areas as infra-red, millimetre waves, and ultrasonics.[59]

Despite both scientific and organizational successes during his year as chief superintendent, Lewis did not achieve all his goals for the establishment. He had hoped to move TRE close to a major university where it could operate as a national electronics establishment serving both civil and defence needs. Instead TRE remained in Malvern, where it continued to serve as an important research establishment.[60]

The war proved to be an important turning point in Lewis's life. He had enjoyed his time at the Cavendish, working with outstanding physicists in a university environment, and remained on leave of absence from the university throughout the war. But unlike many of his colleagues, he chose not to return to university life afterwards. His role as research director at TRE suited him well. It allowed him to exploit his scientific abilities as well as his capacity for synthesizing large amounts of information. These talents, essential for coordinating the many, varied scientific activities with the needs of the armed forces, made him an effective force at TRE. After the war, Lewis's scientific abilities and service to the country were both acknowledged. In March 1945, he was elected as a fellow of the Royal Society; a year later he was appointed a Commander of the British Empire.

The war years had been physically and mentally draining for the radar scientists but Lewis, unlike many of his colleagues, seemed unaffected by the effort. At TRE Lewis developed his style as a research director. In the pressure-filled atmosphere of wartime Britain, his ability to focus on the task at hand was extremely useful. He enjoyed the intensity of the work and the concentration on a specific goal. His talents, which had evolved during the war years, would prove to be essential for tackling the new challenges awaiting him in Canada.

CHAPTER THREE

Chalk River

Rising abruptly up out of the Vale of Evesham, the Malvern hills back the western side of the town of Malvern, final wartime home of the Telecommunications Research Establishment. Famous today for the purity of its spring waters, Malvern's reputation was established during the nineteenth century when a fashionable doctor built a spa there. People flocked to the town to experience the curative powers of the Malvern waters and to walk on the hills. Dominating the town, the hills stretch from north to south and are easily accessible for long rambles. For an inveterate hiker like W.B. Lewis, Malvern must have seemed the ideal spot to pursue the work he so loved.

But in September 1946, Lewis left Malvern and England to start life in a new country. His surroundings would again be beautiful although more rugged. The climate would be very different, with midwinter temperatures dipping to negative forty degrees Celsius. The pressures of war had led to the establishment of an atomic energy laboratory at Chalk River, a remote location in Canada. In 1946 Lewis agreed to become its director. Available documents do not point to any single compelling reason for Lewis's decision to go to Chalk River but a number of factors do emerge.

With the war over, TRE's primary *raison d'être* had vanished. Radar would remain important for both civil aviation and future military applications but the sense of urgency was gone, and this was reflected in the atmosphere at the establishment. The scientists were exhausted. It had been a gruelling war for them, made up of long hours and many triumphs, but also many frustrations. Some scientists chose to remain at TRE. But for most the time had come to return to the universities they had left at the outset of the war. The government pushed for reductions in the civil service. Now the chief superintendent of the establishment, Lewis had to fight to keep it a reasonable size. Indeed, he had to fight to keep the team at Malvern together at all.

Lewis's year as director of TRE proved to be a difficult one. Proposals put forward for the establishment's future had been, by and large, ignored. The constant struggle to maintain the strength of TRE must have been frustrating. In many ways it paralleled Britain's larger effort to rebuild after six devastating years of war. Lewis might well have considered leaving his government post and returning to an academic appointment. But university jobs were scarce and, in some respects, ill suited to Lewis's talents. Lewis had proven himself to be an able experimental physicist during his Cavendish years, but he had discovered through his work on radar that his true forte lay in synthesizing the research efforts of others. At a university, individual research was usually pursued, perhaps in collaboration with one or two colleagues in the same department. That type of situation would not have maximized Lewis's talent for gathering together large amounts of diverse information in pursuit of a single goal. A further drawback of a university appointment was lack of funds. As Lewis once commented to a colleague when asked why he had not accepted a university chair, no university could provide him with the money, human resources, or freedom he had had at TRE, and would have again at Chalk River.[1]

These factors would have been compounded by the depressed state of the laboratory Lewis would likely have returned to – the Cavendish. Since Rutherford's death in 1937, the Cavendish had declined as a centre for nuclear physics. Most of the scientific luminaries who had been there with Lewis during the 1930s had left. By 1947, in the words of one student, the Cavendish was "shabby, rundown and unenthusiastic."[2] Although it would regain its pre-eminence in other scientific fields, it would never again be a centre for nuclear physics. If Lewis sensed this decline, he would likely have looked elsewhere for a job.

This combination of circumstances – the difficulty of maintaining TRE in postwar Britain and the scarcity of jobs in universities – made Lewis receptive to the suggestion that his name be put forward as John Cockcroft's replacement as director of the Atomic Energy Division of the National Research Council of Canada (NRC). To understand the importance of this position, it is helpful to outline the history of the NRC and the development of its Atomic Energy Division during the Second World War.

The National Research Council was established in 1916 to coordinate scientific contributions to the war effort. Originally known as the

Honorary Advisory Council, it was in its early years just that: a council of scientific advisers. In the years after World War I the council worked to encourage scientific and industrial research in Canada. Initially it did not carry out any scientific work itself, but instead provided scholarships to the universities to promote scientific and industrial research. After considerable lobbying during the 1920s, however, the NRC received permission to construct a central complex of laboratories.[3]

The National Research Council laboratories were completed in 1932. Scientists at the laboratories performed industrial research of a long-term nature and provided standards measurements. A program of scholarships and grants continued to encourage basic research at the universities. The effects of the Depression and a scarcity of trained scientists inhibited growth during the 1930s, but after the outbreak of war in 1939, the NRC expanded rapidly. Within a few months, the number of staff increased tenfold.[4]

By early 1940, NRC scientists were deeply involved in war-related research. In that same year, mindful of their possible wartime applications, NRC physicist Dr G.C. Laurence began experiments related to the recent startling developments in nuclear physics.

Since Chadwick's discovery of the neutron in 1932, scientists had been busy investigating the effects of bombarding various elements with neutrons. The results obtained with one element, uranium, had been a source of confusion for some time. This confusion was finally resolved in a paper published in January 1939 by two German radiochemists, Otto Hahn and Fritz Strassmann, which showed that the neutron bombardment of uranium resulted in the formation of elements with approximately half the atomic number of uranium.

This unexpected result was immediately interpreted by Lise Meitner, a German radiochemist, and the Austrian physicist Otto Frisch as the breaking in half of the uranium nucleus. They pointed out that such "fission" would release a very large amount of energy – a fact confirmed experimentally by Frisch a few days later. The fission process also liberates neutrons, which in turn strike other nuclei, releasing more energy and still more neutrons. This is called a "chain reaction."

It was recognized at an early stage that there were in fact two ways of achieving a chain reaction: one using the fission neutrons directly, and the other using a "moderator" to slow the neutrons down. The fast neutron chain reaction can be uncontrolled as in a bomb, or controlled as in a "fast reactor." The slow neutron chain reaction is used in "thermal reactors," so-called because the slowed-down neutrons have energies equal to the thermal agitation energies of the moderator atoms. In both cases, the process results in the release of an enormous amount of energy.

The fast neutron chain reaction cannot be achieved with ordinary uranium but only with the more readily fissile isotope uranium-235 (U-235), whose abundance relative to the principal isotope U-238 is 0.7 percent.[5] On the other hand, a slow neutron chain reaction is possible with ordinary, or "natural," uranium and a suitable moderator. It was clear by December 1940 that such a reaction would be feasible using "heavy water" (water formed from the heavy isotope of hydrogen, deuterium) as moderator. The possibility of using carbon, a less efficient but much more readily available moderator, was not demonstrated until later. The world's first chain-reacting "pile" (or reactor), built by Enrico Fermi and his colleagues at Chicago, went "critical" on 2 December 1942. It was called a pile because it was constructed simply by stacking blocks of graphite drilled out to accommodate the uranium.

A slow neutron chain reaction is a useful source of energy, for example for submarine propulsion and electricity generation, and a source of neutrons for various purposes. But the priority given to its realization arose from its ability to produce plutonium (PU-239) from neutron capture in U-238. Plutonium can be used instead of U-235 to make a bomb, and its chemical separation from uranium is easier than the isotopic separation of U-235 from U-238. While developing the first atomic bomb, the United States Manhattan Project pursued both routes simultaneously.

Fast reactors are used for "breeding" plutonium. The excess neutrons from a plutonium-fuelled chain reaction are captured in uranium to produce plutonium faster that it is being consumed in the chain reaction. This is not possible by use of slow neutrons.

Returning to the situation in Canada in 1940, Laurence was the first person in the world to set up a small (ten-ton) uranium oxide and carbon pile to try to establish whether a larger version could be made critical. The experimental work progressed slowly as Laurence could only do it when other work was not pressing. By the summer of 1942, it was clear that too many neutrons were being lost by unproductive capture in impurities and criticality would not be achievable. Despite their unsuccessful outcome, however, these experiments marked the beginning of atomic energy research in Canada.

Scientists in other countries were also deeply involved in investigations into the new developments in nuclear physics. In France, the work of Marie Curie, *grande dame* of research on radioactivity, had established a tradition of excellence that was carried on by her daughter and son-in-law, Irène and Frédéric Joliot-Curie. A small group, including the Austrian physicist Hans von Halban and Russian-born Lew Kowarski, had been working with the Joliot-Curies at the Collège de France in Paris. Since learning of the discovery of fission they had set about determining the number of neutrons released when a uranium atom split in two. By April 1939 they were able to publish a paper showing that on average 3.5 neutrons were released, which meant that a chain reaction was possible. Their interest now lay in trying to build a nuclear reactor. Early efforts went towards a homogeneous mixture of uranium oxide and water. But by January 1940, they realized that hydrogen absorbed too many neutrons; another moderator would have to be found. Deuterium (a heavy isotope of hydrogen) combined with oxygen to make heavy water. This excellent moderator was very scarce but the French were able to secure the entire world supply from a Norwegian company. By the spring of 1940, experiments using uranium oxide and heavy water were being set up in France.[6]

But world events would soon interfere with the scientific plans of the Collège de France group. In May the scientists fled before the advancing German armies to the south of France. In mid-June, Joliot spoke to both Halban and Kowarski. Surrender was imminent. Although he had decided to remain in France he asked them to carry on their experimental work abroad. Within days Halban, Kowarski, and their precious cargo of heavy water were in England. They wrote up a report of their work in France and waited to see where they would be allowed to continue their research.[7]

In England, fission research had been progressing slowly. Although British scientists had recognized that it might be theoretically possible to construct a bomb by using the fission chain reaction, many were sceptical about its technical feasibility. Chadwick himself, discoverer of the neutron and a brilliant nuclear physicist, doubted whether an explosive chain reaction was possible.[8] A further difficulty in Britain was the lack of manpower. The majority of British scientists had been recruited to work on the more pressing problem of radar. Certain foreign scientists, however, unable to work on radar for security reasons, were free to examine questions about uranium fission. Two of them in particular, Rudolf Peierls and Otto Frisch, were intrigued by the process of nuclear fission. They calculated the amount of uranium-235 (the isotope of uranium most prone to fission) necessary to achieve an explosive chain reaction. Their estimate that only a few kilograms would be needed suddenly made the construction of an atomic bomb more likely.

As a result of the Frisch-Peierls memorandum, action was taken to further British investigations into these questions. A small subcommittee of the Committee for the Scientific Survey of Air Warfare, the Maud committee, was set up to investigate whether an atomic bomb could be built. By the summer of 1940, work had begun on various problems at laboratories in Liverpool, Birmingham, and Cambridge. The Maud committee remained uncertain whether the group from France would be useful to the British program. The French work, using a homogeneous mixture of uranium, heavy water, and thermal neutrons to try to produce a chain reaction, was far removed from the British efforts towards building a bomb with uranium-235 and fast neutrons. But the French group's interest in constructing a power reactor, or "boiler" as it was then known, had possible wartime applications. The committee decided that the scientists would remain at the Cavendish where they had been temporarily placed.[9]

In the fall of 1940, the Tizard mission had crossed the Atlantic to exchange scientific information primarily with American but also with Canadian scientists. The discussions related mainly to radar developments but the British scientists used the opportunity to talk about advances in nuclear research. They visited a number of different laboratories, including George Laurence's at the National Research Council in Ottawa. Cockcroft, a member of the mission, was impressed by Laurence's work and upon returning to England, arranged for a British company to contribute five thousand dollars to this research.[10]

By the spring of 1941, the atomic bomb was no longer merely a subject for scientific pondering; it was clear that, with enough concentrated effort, it could be built. The details were stated in a report drafted by the Maud committee in England. During the summer, a copy of the Maud report crossed the Atlantic. It had a strong impact in the United States, leading to a reorganization of the American bomb project. But it was only after the Japanese attack on Pearl Harbor brought the Americans into the war that their effort accelerated rapidly. During these final months of 1941, the British missed the opportunity of securing close cooperation with the United States on this vital scientific project. Suggestions put forward by the Americans were not acted upon immediately; by the time the British realized that closer integration of the two projects would benefit them greatly, the American effort had far surpassed that of the British and Washington was no longer interested in integration.[11]

During the winter of 1941–42, Halban toured the US project and reported back to the British that the Americans were far advanced. One of the few areas where the British were still ahead was in the work done by Halban's group on a slow neutron system. In July 1941, the

Maud committee had recommended that the group be transferred to the United States. At that time it had been argued that, although the team's investigations had potential future significance, they lacked immediate relevance for the development of the bomb. Given the limits on British manpower and resources, it seemed unwise to expend them in this manner. Now, in an effort to maintain some form of collaboration with the Americans, a number of suggestions were put forward in Britain concerning what to do with Halban's group. Because of its foreign makeup, the American authorities were leery of setting up the team in the United States. Another suggestion, that Halban and a few others join the Chicago team, was not popular with Halban, who stood to lose his position as leader. The third possibility, and the one championed by Halban, was that the entire team be transferred to Canada where they could continue their work on a heavy water pile. In Canada they would be close to American supplies but could retain their independence.[12]

Authorities in Britain concurred and in August 1942 the Canadians, with surprisingly few questions, agreed to house the project. It was decided that the best place for the scientists to work would be Montreal. Supplies would be more readily available there than in the capital city, and the cosmopolitan nature of Montreal would make the influx of foreign scientists less conspicuous, thus helping to keep the existence of the laboratory a secret. What remained remarkably unclear was the exact mandate of the "Montreal laboratory," as Halban's group came to be known. The Canadians, anxious to help out their British allies, did not ponder even the immediate objective let alone what would become of the laboratory in peacetime.[13]

The arduous task of transporting personnel and equipment across the Atlantic was soon underway. But problems arose when the British started asking the Americans for the information and supplies that were necessary for the laboratory to function. Over the next few months cooperation between the two countries became severely strained. The Americans decided to restrict cooperation to what would be useful to them. This decision, communicated in January 1943 to C.J. Mackenzie, president of the National Research Council, slowed the work of the Montreal laboratory to a near standstill while the three governments tried to work out a compromise.[14]

A compromise was finally reached and incorporated into the Quebec agreement of August 1943. The British announced their desire to help solely in the military sphere, dispelling American fears about Britain's interest in postwar commercial applications of nuclear power. There would be full cooperation when both countries were working on the same problems. After further negotiations it was agreed that the Montreal laboratory in Canada could stay but that the director, Halban, must go. Halban's arrogance had alienated many of his staff, the NRC president, and, most importantly, officials in the American program. His replacement, John Cockcroft of Cavendish and radar fame, arrived in April 1944. His immediate task was to revitalize the laboratory.

After Cockcroft took over the leadership of the group, events moved quickly. By August, a permanent site for the new laboratory had been chosen. Up the Ottawa River, two hundred kilometres north-west of Ottawa, the new Chalk River project got underway. Considering the political problems the group had been forced to endure, the remoteness of the site, and the difficulties of procuring both personnel and supplies during wartime, progress was remarkably rapid. By September 1945 a zero-energy experimental pile (ZEEP) had gone critical – the first reactor to do so outside of the United States. Construction of the larger reactor – known as NRX for National Research Experimental – continued apace. However, the war had ended in August and changes were bound to come.

The most important of these for Chalk River was to be a change of director. John Cockcroft had left his radar work during the war to take over the leadership of the Montreal laboratory. When he first arrived, the morale of the scientists was at rock bottom. Under Cockcroft's direction the group had accomplished a great deal and established itself as an excellent nuclear laboratory. The NRX reactor, designed under Cockcroft's guidance, was equipped with many outstanding experimental facilities. These were Cockcroft's legacies to the Canadian project. But the British were anxious to set up their own nuclear establishment to work towards the construction of experimental reactors, plutonium-producing reactors and possibly electricity-generating reactors. They wanted Cockcroft to head this laboratory; once again, Cockcroft responded to the patriotic call and prepared to return to England.

Once it was clear that Cockcroft would not remain in Canada the question of his replacement arose. Both NRC president C.J. Mackenzie and his minister, C.D. Howe, were annoyed at the sudden announcement that Cockcroft would direct the new British facility; Howe first learned the news from his morning newspaper. Their fears extended beyond the loss of Cockcroft. Many of the staff at Chalk River were British and both men worried that the abrupt removal of the British scientists could jeopardize the Canadian project. The question of staffing the project would remain uncertain for the time being. What was certain was that the new director would be chosen by Howe and Mackenzie and would be a Canadian. The problem was that there was no large pool of Canadian scientists to choose from. In fact there were remarkably few who Mackenzie felt would be suitable for the job. One Canadian candidate, Walter Zinn, had received his doctorate from Columbia University and continued working in America, where he had made substantial contributions to the Manhattan Project and been in charge of the United States heavy water reactor. But his refusal, and that of another Canadian, R.L. Thornton, left the NRC president more receptive to suggestions from his British friends.[15]

Cockcroft felt quite strongly that his replacement should not be an American since, as he put it, "we desire to retain the closest ties with the Chalk River laboratories."[16] The British already had a replacement in mind: W.B. Lewis. Lewis came with the highest of recommendations. Both Cockcroft and Chadwick praised his technical and administrative talents.[17] Howe and Mackenzie, however, were still unwilling to follow without question the suggestion of the British. But the British were determined to have Lewis accepted and were willing to go to great lengths to succeed. In March 1946 it was suggested that Sir Stafford Cripps, the president of the Board of Trade, write a personal letter to Howe recommending Lewis on the basis of a personal knowledge of Lewis's work at TRE. If this failed to convince Howe and Mackenzie, then Prime Minister Clement Attlee would consider writing to Prime Minister Mackenzie King in the hope that this would swing the balance.[18]

The fact that the British were even considering going to such lengths to secure the position for Lewis underscores the importance they placed on having a director sympathetic to British needs at Chalk River. It also suggests that they were trying to make amends for their unilateral decisions of the previous fall. Cockcroft recognized that it would be a long time before facilities as advanced as those at Chalk River would be available in England. Much crucial scientific and technical information could be gained through close collaboration between the two projects in the upcoming years. Clearly, Cockcroft counted upon Lewis retaining a certain amount of British allegiance in his position as the new "Canadian" director.

In the end such strong measures were unnecessary. Once Zinn and Thornton had refused the position, Howe and Mackenzie were quite willing to consider Lewis. Mackenzie already knew Lewis and was aware of the work he had done at the radar establishment. Having confirmed with the British authorities that some British scientists would remain temporarily at Chalk River, Mackenzie was quite happy to accept British advice. By July negotiations had been completed. The new director of NRC's Atomic Energy Division arrived in Canada in September 1946 to take up his position.

CHALK RIVER

There is no known record of Lewis's first impressions of the plant he would direct for the next twenty-seven years. Certainly, as an Englishman accustomed to the British countryside, he must have been impressed by its remote location and wild surroundings. The plant road, which branches off from the main highway near the small town of Chalk River, winds by lakes and through forests. The surroundings are untamed and startlingly beautiful. After driving over a final ridge, the plant comes into view. Behind the buildings lies the river with a mist slowly rising up from it. Across the river, pines meet solid rock face. The juxtaposition of this wilderness with the plant is remarkable. Canada's most scientifically ambitious and expensive research effort was growing steadily in the middle of the Canadian bush.

On the left, after passing through a security gate, was the Administration building, which housed Lewis's office. Constructed from the familiar army-type white clapboard, these buildings have remained in place to the present, a reminder of the plant's military origins. Continuing down the main plant road, Lewis would go through another security gate into the "active" area where operations involving high-radiation fields were carried out. It was within this area that the new pile was being constructed when he arrived.

What was surprisingly unclear when Lewis took up his duties was the purpose of the plant, now that the war had ended. In 1944 the Americans had agreed to support the project with necessary supplies because it was centred around construction of a pilot heavy-water natural-uranium pile with plutonium-production potential. But the war had ended in 1945, and it would be almost another year after Lewis's arrival before the pile went critical.

Lewis appears to have received little direction concerning the future of the project from his superiors, C.J. Mackenzie and C.D. Howe. They were aware that the Canadian atomic project was more expensive than all other Canadian research projects combined. And yet its future still remained uncertain. The possibility of moving towards bomb production would have been rejected as politically unpopular – if it was ever seriously considered at all. Power production was a possible future goal but Canada's abundant supply of hydroelectricity did not make it an immediate priority. It would appear that Howe and Mackenzie sensed in a vague way that there was a future in atomic energy and if Canada was to be involved it would be through Chalk River. For this reason alone it was worthwhile continuing to support the laboratory.[19]

The situation in Canada was not unique. Cockcroft's newly established laboratory, Harwell, had only vague terms of reference in the early postwar years. It was largely left to Cockcroft to determine the function of the laboratory. Similarly with Lewis at Chalk River. Both countries appeared ready to follow the example already set by the American research establishments: pursuit of any kind of research connected with atomic energy.[20]

Canada had one important advantage, however. NRX had been designed to be a first-class research reactor and, once completed in 1947, it amply fulfilled its promise – indeed, for many years it was simply the best research reactor in the world. Lewis's first responsibility was to see that its capabilities were fully exploited, for both pure and applied research. Then from the results of this research and, no less important, from the operational experience gained, a Canadian atomic energy program could take shape.

The NRC division Lewis took over in September 1946 was already well established. The programs and staff from the Montreal laboratories had all been transferred by July. The overriding priority of the new plant was to complete and start operating NRX, which was badly behind schedule. This, along with the management of the Chalk River site and the village of Deep River (see chapter 4), was the responsibility of Defence Industries Limited (DIL), a large Crown corporation set up during the war to build and run munitions factories. By 1946, DIL wanted to terminate its activities and, in February 1947, NRC took over the whole project, on behalf of the Atomic Energy Control Board (AECB).[21] There were three divisions: Lewis's, which was renamed Atomic Energy Research; Engineering, under Ken Tupper (previously Lewis's deputy director); and Health, under Dr W.E. Park.

The new organization, with the engineering division responsible for NRX, still left Lewis time to acquaint himself with the scientists under his direction, the facilities in place and under construction, and research projects that were underway. Furthermore, "research" was given a very broad interpretation and Lewis's role in charting the future of the Chalk River project would be central. Indeed, when Tupper left in March 1949, Lewis took over all his responsibilities as well as retaining those of research director.

By the time he arrived at Chalk River, Lewis had already developed a distinct managerial style for running a scientific research establishment. Building on his experience at the Telecommunications Research Establishment, Lewis would only loosely follow the line organization pattern. If he needed to know the results of an experiment, he talked directly to the scientist in charge, bypassing branch heads and other senior staff. This approach likely stemmed both from his training at the Cavendish Laboratories and from his time spent at TRE. At the Cavendish, ideas were discussed freely among scientists from many different research groups while at TRE, with the pressures of war weighing heavily upon the scientists, there was no time to follow a strict chain of command. This method of direction also suited Lewis's personality and abilities. His fascination with every type of research being performed at the plant, combined with his ability to synthesise rapidly large amounts of diverse information, made Lewis particularly well suited for his job as research director.

When he arrived in 1946, Lewis had a great deal of catching up to do in a field he had been completely divorced from during the war. A prodigious reader, he quickly acquainted himself with the advances in atomic energy that had occurred during the war and remained abreast of the literature as it was produced in the postwar years. Within the plant, memos and reports kept him informed on every aspect of ongoing research. His phenomenal memory is recalled by everybody who worked with him. At times it appeared that Lewis was more knowledgeable about the experiment in question than the scientists reporting on it. Lewis also used committee meetings to gather information about the work underway at the plant. Scientists around the table provided the information asked for and Lewis, upon considering the facts, made a decision if one was necessary. Finally, the director made it his personal responsibility to vet every paper published out of Chalk River. Before leaving for a conference, scientists were asked to practise delivering their paper to an audience. Invariably Lewis was the toughest critic. His standards were high and this ritual placed a great deal of pressure on the scientists, but Chalk River quickly gained a high reputation in the international scientific community.

The selection of research topics was an important part of Lewis's job as director. Lewis gave his scientists a great deal of freedom in choosing research projects. He trusted them to choose useful and interesting areas to investigate. With his quick mind and enormous appetite for work, he was able to follow their progress and, in discussions, challenge them on aspects of it. Scientists in certain disciplines, most notably chemistry and biology, would at times complain that he did not really understand their areas of expertise. This was doubtless true; but to a remarkable degree, Lewis did remain closely in tune with the enormous variety of scientific research performed at Chalk River.[22]

In addition, certain administrative duties required Lewis's attention. The problem of recruiting and retaining staff, for example, acquired special urgency in the postwar years, when the lure of larger salaries in industry and in the United States attracted promising young Canadian scientists. By the fall of 1947 a number of the research staff at Chalk River felt that the problem had become acute. In a letter to NRC president C.J. Mackenzie, a group of scientists from Chalk River complained that the departure of many British-paid staff was adversely affecting their research.[23] Mackenzie adamantly refused to admit that the scientists at Chalk River were underpaid and dismissed the argument that inadequate financial incentive had been offered to men like Zinn. Instead, Mackenzie pointedly observed that "where there is first-class, enthusiastic scientific leadership by the Directors and Section Heads, there is the best chance of getting and maintaining staff."[24] Lewis's views on this question are unknown but he likely sided with the senior scientists. Recognizing that retaining high-quality scientists was crucial to Chalk River's success, Lewis was closely involved in recruiting new people and worked tirelessly to advance their subsequent careers.

One group that Lewis did not approach were women scientists. There were few women professionals at Chalk River in the early days and those hired were not swiftly promoted. Lewis did not establish this precedent, but he did nothing to change it. As one woman scientist commented, it is unlikely that he ever thought about it.[25]

Although Lewis did remain closely involved with hiring decisions, Mackenzie tried to limit Lewis's responsibilities for other administrative problems. As he wrote to Cockcroft early in 1947, he wanted "to avoid having Lewis concerned with a hundred and one administrative details about the plant as I need his concentrated thinking and attention on the purely research and scientific aspects."[26] Mackenzie need not have worried about Lewis's involvement in the research side of the plant. Lewis was, and remained throughout his career, fascinated with the scientific research performed at Chalk River.

RESEARCH

In the early years after Lewis's arrival at Chalk River, much of the work performed was pure, as opposed to applied, research; that is, research performed simply to increase knowledge on a particular subject, "undertaken without a specific application in mind at the time."[27] It should be noted, however, that in those early years, the distinction between pure and applied research was often unclear. During this period, "pure" scientists designed shielding, built their own electronics, and blew their own glass. Later, as the applied programs at Chalk River developed, the distinction between the pure and applied scientists would become more defined. But because applications often evolve out of pure research and applied research can lead to fundamental discoveries, the two areas can never be regarded as completely distinct.

Pure research would always remain important at Chalk River. In large part this can be attributed to Lewis's continued support and enthusiasm for fundamental research throughout his career there. His involvement in pure research programs peaked in those early years; later, he would become increasingly occupied with the applied side of the power reactor program. Thus, a discussion of Lewis's role must necessarily focus on the early postwar years. This is not meant to downplay the importance of later research but rather to reflect Lewis's own shifting focus of interest.

The NRX reactor was central to the research performed at Chalk River. Although it was originally planned as a pilot plutonium-producing pile, NRX was designed as a first-class research reactor with a high flux and equipped with excellent facilities. After its initial start-up in July 1947, its performance proved even better than had been predicted. Although an accident at the end of 1947 caused the reactor to be shut down, by February 1948 it was back in action. By May the reactor was operating at the ten-megawatt (MW) level and reached the design value of twenty MW in January 1949. During 1948 the maximum flux of 4.3×10^{13} neutrons per square centimetre was reached, which was believed to be, as Lewis proudly recorded, "the highest steady flux available for experimental purposes."[28]

The scientists at Chalk River did not hesitate to exploit the research facilities of this reactor. Much of the work that the nuclear physics branch, headed by B.W. Sargent, performed during the early years was devoted to a basic understanding of how the pile worked. Extensive studies were made of the behaviour, shielding, and neutron fluxes of NRX, as well as the fission process itself.[29] Notable among the early experiments were careful studies of the neutron. In 1948, two Chalk River physicists, R.E. Bell and L.G. Elliott, made a significant correction to the previously accepted value of the mass of the neutron by measuring the energy of the gamma ray that is produced when a proton captures a neutron. Two years later came J.M. Robson's classic measurements of the lifetime of a free neutron and of the energy of its decay electrons.[30]

The most important contribution of the physics group was made by the team working with B.N. Brockhouse. Their development of the neutron-scattering technique has been crucial to the development of condensed matter physics. It could in fact be argued that this was the most important Canadian contribution to physics.[31]

Although the NRX reactor was their main research tool, the physicists also used a three-million-volt Van de Graaff generator for experimental work. Bombarding atomic nuclei with the accelerated ions from this machine led to their "artificial disintegration" – a widely used technique for the study of nuclear structure. The accelerator group would remain strong at Chalk River throughout Lewis's tenure as research director. In 1959, a much more powerful accelerator was added to the laboratory, the world's first "tandem" Van de Graaff, and this machine was replaced with a larger tandem in 1967. Lewis's continued support of this branch of physics is an example of his commitment to pure research at Chalk River.

Lewis had trained as a physicist and therefore remained closely involved with the work of the physics branch. It has been suggested that other areas of research suffered as a result of Lewis's central interest in physics. While there may be some truth to this, the fact remains that a great deal of important research was performed outside the physics branch.

The chemistry branch, headed by L.G. Cook, carried out research that straddled the pure and applied categories. A Canadian, Les Cook trained in Berlin and had crossed paths with Lewis at the Cavendish. In 1944 he joined the Montreal laboratory and by 1946, when Lewis arrived, he was chemistry branch head.

The chemistry branch underwent a period of transition during the early postwar years. It still contained many British scientists hard at work on projects that were of greater importance to the British than to the Canadians. In 1948, for example, one British team was developing rapid methods for the separation of individual fission products from irradiated natural uranium. As will be seen, such use by the British of Chalk River facilities caused some controversy between Britons and Canadians.

Most of the work performed in this branch, however, was of great importance to the Chalk River project. The effect of neutron irradiation on solids was measured to gain a better understanding of some of the properties of solids. The chemical response of materials under intense irradiation was examined to gain a clearer picture of the composition of the materials. This information was also used to aid the engineering branch in designing future reactors. This is a good example of pure research finding an immediate application.[32]

Other experiments by chemists exploited the excellent research potential of the NRX pile. Its exceptionally high flux made it possible to determine some nuclear properties of isotopes, which would be impossible to do elsewhere. Investigations performed in collaboration with the nuclear physics branch examined the nuclear and chemical properties of synthetic elements (plutonium, americium, and curium) derived from uranium. Other fundamental studies included work on the decomposition of water under irradiation and investigations of the chemistry of uranium compounds.

In 1949, the chemistry and engineering branches were brought together under one subdivision. This move recognized the important contribution of chemical work towards the engineering of future reactors. George Laurence became assistant director in charge of the new chemistry and engineering subdivision. This same reorganization affected the final area of research under Lewis's direct supervision by creating an enlarged biology and health radiation subdivision. Directed by A.J. Cipriani, this group brought together work performed in different branches concerned with various problems in radiobiology and the protection of the health of personnel.

The biology branch at Chalk River had existed since the days of the Montreal laboratories. The program set out to study three broad areas: the mechanism of action of various types of radiations on cells, chemical mechanisms of radiation effects, and the genetic effects of radiation. These were all ambitious areas of research, aspects of which have continued into the present.

As in other divisions, the work of this branch was often a mixture both of different disciplines and of fundamental and applied research. After the completion of the NRX reactor in 1947, the scientists quickly began to exploit the ready supply of radioactive tracers. Other work was directly related to the safety of the workers at the plant. The toxicity of different radionuclides was carefully measured to determine the allowable limits of radiation exposure. Programs were established to enhance the protection of project personnel against exposure to radiation and methods were developed for the control and removal of radioactive contamination.[33]

Lewis's involvement in the area of biological research was not extensive. When there were no problems with the ongoing research there would be little interference from the director, but if problems arose he would be at the branch asking questions. One area Lewis found particularly interesting was health physics. With his own extensive knowledge of electronic counting devices, Lewis understood this area, which dealt with instruments and measures of external dosimetry, better than those of pure biology. It is likely that this flawed understanding of biology affected his treatment of the branch when funding decisions had to be made. There was a feeling within the biology branch that it was treated like a "poor cousin" in the establishment. Part of the blame for this must be laid at Lewis's door. Lewis was always interested in the work of the branch and doubtless believed its research to be extremely important, but it appeared that he had a certain idea of how large the branch should be and would not fight for its expansion.[34] In this branch, as in chemistry, it was believed that Lewis understood and supported physics to a greater extent.

Under Lewis, and even after his departure, basic research at Chalk River was maintained at a fairly steady level. The programs changed, of course, and became more narrowly focused as funds were cut back. Some of the basic research, especially in chemistry and materials science, was closely connected to the applied-research program, and the biological research was essential in a number of ways. Lewis described the importance of fundamental research in terms of its ability "to provide a base, and sometimes a novel and revolutionary base, on which to build a technological advance that by meeting the competition yields an economic benefit."[35] It has to be recognized, however, that the basic research produced little that was of direct value to the power reactor program.

Basic research dominated in the early days at Chalk River but applied research grew steadily and eventually surpassed it, spanning an enormous range of activities. Lewis's involvement here was close, closer than with pure research, and it would be a Herculean task to provide even a summary of all the applied research undertaken during his tenure. Accordingly, the rest of this section gives a very brief account of just those areas where Lewis's interest in the research, and his contributions to it, were particularly noteworthy.

Lewis was chairman of the NRX Pile Operating Committee, which brought research and operations staff together and gave Lewis first-hand knowledge and experience. Special mention should be made of the in-reactor "loops," so called because of their closed-circuit coolant systems, in which fuel and other materials could be tested under intense neutron bombardment and at controllable temperatures. The United States loop experiments, which turned out to have enormous significance for the Canadian power program, began in 1951.

Understanding the operation of NRX was of great importance from many points of view, and the solutions to the problems that arose – e.g., moderator chemistry, corrosion, fuel distribution – helped pave the way to the succession of reactors described in following chapters. Of particular interest to Lewis was the decrease in reactivity of fuel rods as the irradiation proceeded. This decrease set the "burn-up" i.e., how much energy can be extracted from a rod before it has to be removed from the reactor because it is absorbing too many neutrons unproductively. NRX itself was used to measure these reactivity changes, and one notable experiment resulted in serious damage to the reactor in December 1952 (see chapter 5).

Burn-up is the fundamental determinant of fuelling cost, and it depends on the details of neutron production and absorption. Neutrons are the currency of nuclear power and their unproductive waste was anathema to Lewis. As future chapters will show, all reactor concepts were subjected to his inflexible insistence on "neutron economy." Lewis devoted a great deal of time and effort to ensuring neutron economy, performing elaborate calculations of expected burn-up and supervising an increasingly wide range of experimental programs aimed at providing more complete and accurate input data for the calculations. These programs involved the construction of two more small reactors: ZED-2, which supplemented ZEEP for measuring the reactivity of different spatial arrangements of fuel, and PTR, which measured the reactivity of irradiated material. Lewis also directed a large program that provided detailed chemical and isotopic analyses of samples of irradiated fuel. To provide a clearing house for all these data and a forum for evaluation and planning, Lewis set up a Nuclear Data Committee in 1954. He always chaired this committee and its final, eighty-second meeting took place nine days before he retired in June 1973.

Given his recognition of the value of neutrons ($1,500 a gram in 1952), it is not surprising that Lewis considered other sources of neutrons that could play a useful role in fission-based energy systems. In a couple of 1952 reports, he discussed the production of neutrons by the high-energy proton bombardment of heavy elements.[36] An experiment was set up to measure the neutron yields from different targets, using the high-energy protons present in cosmic rays, at a high altitude (11,000 feet) at Echo Lake in Colorado. This experiment was a major logistical undertaking and technically quite difficult. It was feasible only because, at that time, the Chalk River physics program was broad enough to include an active program in cosmic ray research.

This so-called "electronuclear breeding" has not so far proved to be economically viable, but it was this method of neutron production that would have been used in the Intense Neutron Generator, discussed in chapter 7. During the ING period Lewis devoted considerable effort and ingenuity to devising novel schemes for "factory accelerators" that would make electronuclear breeding economic.

In a 1952 paper entitled "Atomic energy objectives for applied nuclear physics research," Lewis also recognized another possible source of useful neutrons, thermonuclear fusion, but only as a distant prospect. By 1958, following the declassification of previously secret work, success seemed much closer, but Lewis remained sceptical.

Because of the difficulty of achieving controlled thermonuclear fusion, he believed that it was more likely to be economically viable as a source of neutrons feeding a near-breeding fission system than as a stand-alone power source.[37] Accordingly he was content to await events, and time has certainly validated his caution.

The development of the CANDU nuclear power reactor was the major applied research project at Chalk River. The decisions that the CANDU system would be based on the use of uranium oxide fuel in pressure tubes were reached by Lewis with considerable reluctance. Both uranium oxide and pressure tubes offended against the principle of neutron economy: oxide was vastly inferior to metal as a fuel except in the absolutely vital matters of resistance to corrosion and deformation, and the use of a pressure vessel surrounding the whole reactor would have removed a great mass of necessarily neutron-absorbing material (i.e., the pressure tubes) from the inside of the reactor. However, the construction of a pressure vessel for a full-scale CANDU was considered to be totally impracticable.

Regarding the fuel, Lewis started with a clear enunciation of two criteria that had to be met and assembled a strong team to tackle what was in fact a very complicated system. A particularly tricky aspect was whether the release of gaseous fission products would be sufficient to distend the thin sheath used to cover the fuel (this for reasons of neutron economy, of course), but it turned out to be a manageable problem. Very satisfactory fuel was developed in time for the power demonstration reactor, NPD, but work continued for many years to improve reliability. CANDU reactor fuel can fairly be described as an unqualified success.

The pressure tube story is not a happy one, however. The US-developed material Zircalloy-2 was chosen for NPD, but already in 1958 Lewis ordered work started on a stronger material, Zr-2.5 % Nb, on the basis of information from the Soviet Union. By 1966 ZrNb pressure tubes had been developed to the point where they were chosen, over Zircalloy ones, for Pickering 3 and later reactors.

The subsequent developments – unexpectedly rapid creep and growth, delayed hydrogen cracking at improperly rolled joints, and the 1983 failure of a Zircalloy pressure tube that had made contact with a calandria tube in Pickering 2 – all occurred after Lewis retired. It is difficult to attach any blame for the last two items, which were in the nature of unforeseeable accidents, to deficiencies in the R&D program, at least when it was still under Lewis's control. The investigation of creep and growth, which was known to be a potentially serious problem, was limited by the relative inadequacy of the irradiation facilities. In any case it is an economic rather than a safety problem. Regarding the possibility of pressure-tube failure and its consequences, an extensive test program had been carried out with deliberately defective tubes and the results had been completely reassuring – a crack would always be detected, by leakage, before it grew to critical length and the tube failed.

Lewis was well aware of the limitations of scientific understanding. He described the situation in the following words: "It has been remarked that the Bronze Age metallurgists passed on their practices and traditions without understanding the materials with which they worked. The same is true today of our use of materials at high temperatures and in reactors. The bulk of the scientific understanding is yet to come for the processes of creep, corrosion, embrittlement, fracture, swelling, distortion and local changes of position."[38]

POLICY

Lewis's involvement with research at Chalk River was an important part of his job that brought him into direct contact with the scientists working for him. The other main aspect of his job concerned planning for the future of the laboratories. In making these decisions Lewis worked closely with the president of the NRC, C.J. Mackenzie.

Chalmers Jack Mackenzie was born in New Brunswick in 1888. In 1909 he graduated with a degree in engineering from Dalhousie University. One of his courses had been taught by the young C.D. Howe, a man who would later figure prominently in his life. After completing his degree, Mackenzie travelled west where work was more plentiful. Within a few years he was offered a teaching position at the University of Saskatchewan. He would remain closely associated with the university as it developed a complete engineering school. In 1935, Mackenzie became a member of the Honorary Advisory Council of the National Research Council; four years later he was asked by NRC president A.G.L. McNaughton to become acting president. In 1944 McNaughton formally stepped down and was succeeded by C.J. Mackenzie.

As acting president of the NRC during the Second World War, Mackenzie was in charge of an enormous research effort of which the atomic energy project, first at Montreal and then at Chalk River, was only a part. Mackenzie prided himself on his administrative talents, and early accounts of his career reflect this view. But some historians have been critical, describing Mackenzie as a poor administrator who was "unable to divest himself of the day-to-day detail and concentrate on larger questions of policy."[39] Mackenzie's statement about avoiding having Lewis concerned with "a hundred and one administrative details" suggests that he hoped to maintain his hands-on approach in the postwar years. He continued to believe he could successfully manage both the NRC and its atomic energy division in this manner despite the latter's growing size and complexity. Lewis accepted Mackenzie's view that he (Lewis) should be in charge of the scientific direction of the plant, but made it clear that he too would have a clear say in the administration and policy decisions concerning Chalk River.

Lewis's experiences at TRE during the war had shaped his views of the relationship that should exist between the scientist and "headquarters." The urgent need for the rapid development of radar during World War II had brought the varying levels of the military and scientific communities into close contact. No longer did orders from the top tell scientists what to invent; instead, scientists and military personnel discussed what had been developed and what was needed. Lewis experienced first hand the value of the scientist's input into policy decisions on the type of radar systems that should be emphasized. This method of interaction was extremely successful during the war and Lewis wanted it carried over to his new position in Canada.[40]

One important policy question Mackenzie and Lewis did not agree upon was the amount of cooperation and exchange that should occur between the British and the Canadians. Lewis approached the question as a scientist, recognizing that collaboration would be to the benefit of both projects. Mackenzie, on the other hand, still smarting from the abrupt removal of Cockcroft by the British, wanted to ensure that the British were not taking advantage of Canadian generosity. For his part, Cockcroft was anxious to establish close relations between Chalk River and Harwell. Cockcroft realized that it would be some time before the British had facilities comparable to those already existing at Chalk River. Of particular importance to the British was access to the NRX reactor when it was finished. Only by performing experimental work using the new reactor could the British answer the many questions involved in building their own.[41]

The British desire to collaborate with the Canadians was heightened by their realization that cooperation with the United States was all but over. In the year after war's end, political controversy surrounded atomic energy legislation in the United States. The first bill proposed had dismayed scientists, who decried the high level of military involvement in atomic energy affairs. It was replaced by the McMahon Bill, which recommended a civilian Atomic Energy Commission; the price for the civilian control was strict provisions concerning the dissemination of atomic information.[42] The passage of this legislation in 1946 effectively ended collaboration between the American and British atomic energy projects. Britain was still able to obtain information from Chalk River, however, which was being supplied with heavy water and uranium by the Americans. The change in US policy made this connection vital to the fledgling British project.

It was a connection that Lewis was happy to maintain; exchange of information, he recognized, would benefit both sides. Because of his involvement in radar throughout the war, Lewis did not have the background knowledge of atomic energy developments that Cockcroft possessed. In taking over direction of the Chalk River plant, Lewis undoubtedly benefited from discussions with Cockcroft. Scientific interchange was, after all, an important part of the scientific process. In collaborating with Cockcroft, Lewis was simply acting as a research scientist.

Nevertheless, Mackenzie firmly believed that the British removal of Cockcroft from the Canadian project had cancelled agreements between the two countries reached during the war. In a letter written to Lewis in February 1947, he dampened Lewis's enthusiasm for offering facilities to the British and stressed that no concrete commitment to bilateral cooperation existed.[43] Mackenzie was concerned that Lewis would go out of his way to help Cockcroft, perhaps to the detriment of the Canadian project. Although the new director was British born and raised, Mackenzie was determined that he would be a *Canadian* director, putting Canada's interests first. As it turned out, Mackenzie had little to worry about. Lewis tackled his new job in the same single-minded manner as he had approached radar and was loyal to the project.

Cockcroft faced a challenging task when he returned to England in the fall of 1946 to direct Harwell, Britain's atomic energy research establishment. Harwell's mandate was very broad: "to pursue atomic energy research in all its aspects."[44] But in the postwar years Britain's immediate purpose was to produce fissile material for a bomb, which meant that Harwell also had to provide information on the chemistry and manufacture of plutonium as well as the peaceful uses of atomic energy, radioisotopes, and, in the future, power. Britain's lack of facilities and the restrictive nature of the McMahon Act meant that Cockcroft had to rely heavily on the Canadian project for atomic information.

Cockcroft tried to formulate a clear idea of what information the British needed from the Canadian project. Most pressing were the data necessary to move ahead with the design of a chemical-extraction plant for the British 300,000-kw pile. Cockcroft suggested that most of the research and development work would be needed by the Canadians for their own chemical-extraction plant for the NRX pile. Both the British and Canadians needed information on the removal of the sheaths – the metallic covering of the uranium fuel – and the subsequent dissolving of the uranium fuel elements. Cockcroft hoped that this information would be available by the summer of 1947, but he planned to leave British chemists and engineers in Canada "for some considerable time" as well as provide additional staff if necessary.[45]

Although Lewis was ready to help the British with both general problems and those of special interest to their program, he first checked that he had Mackenzie's backing. Mackenzie agreed but reiterated his belief that the partnership that had existed during the war had ended abruptly when the British decided to withdraw Cockcroft. Since that time it had been unclear whether British scientists on the project were working for Chalk River or Harwell. Acknowledging that this situation was to be expected, Mackenzie nevertheless indicated that in the future, research requests should come through formal channels.[46]

To a certain extent Mackenzie's wishes were obeyed. In the summer of 1947, Lewis was asked by the British to undertake a series of experiments to determine the effect of neutron irradiation on the oxidation of graphite. This work was judged to be of prime importance to the British production program. Lewis responded favourably to the request but suggested that if the matter was of such importance, it should perhaps be channelled to Mackenzie and, if necessary, the Atomic Energy Control Board.[47] Once formal approval had been received, the experiments were given high priority on both sides of the Atlantic, with telegrams racing back and forth containing requests and results. Also of interest to the British was the phenomenon of blistering. When a uranium rod was inserted in the pile blisters sometimes formed on the surface. These blisters could block water-coolant channels, so that solving the problem was crucial to the development of a water-cooled reactor. The British were particularly concerned that this phenomenon would be accelerated in high-temperature, high-flux reactors.[48] They relied on measurements by the Canadians to resolve these problems.

Less-formal levels of exchange also existed. Shortly after Lewis had succeeded him as director at Chalk River, Cockcroft sent a newsy letter telling Lewis of advances in the British program. He added that he hoped they would be able to swap newsletters as there was much lacking in the picture obtained from official documents.[49] Lewis was glad to oblige. Letters were exchanged across the Atlantic between not only the two directors but also branch heads. Cockcroft, a man known for his brevity of speech, wrote long letters to Lewis – the longest he wrote to anyone.[50]

Interaction with the American projects was more formal. Most decisions about collaboration were decided by the Combined Policy Committee. This body, set up as part of the Quebec agreement of August 1943, facilitated interchange of information among the three countries. The committee was made up of three Americans, two Britons, and one Canadian, C.D. Howe. At a meeting held in December 1947 an agreement on exchange of technical information was drawn up. Nine areas were listed where cooperation was possible, including fundamental properties of reactor materials, research uses of radioisotopes, and the design of natural-uranium power reactors.[51]

The Americans handled interchange of information through a liaison officer working at the plant. The first such officer, Colonel Curtis Nelson, was transferred by the United States Atomic Energy Commission (USAEC) to Chalk River in 1947. Most of the information shared during this period was on heavy water, an area where the Americans, with a far larger outfit, were more knowledgeable than the Canadians.[52] But in other areas the Canadians could offer something in exchange. In 1949, when Mackenzie offered the Americans some fuel rods that had been irradiated longer than was possible in other reactors, they responded immediately by quickly ordering a further eight rods and paying for them.[53]

As director of the Chalk River project Lewis played an important role in these interchanges. He was fully aware of and interested in the experiments undertaken by the Americans and the British using the Canadian reactor. The significance of this collaboration was more than simply the exchange of information. Canadian possession of an excellent research reactor gave the country a new importance in the Atlantic triangle. The series of wartime events that had led to Canadian involvement in atomic energy research had catapulted the country into the atomic club. But in order to stay at the forefront of atomic research, the Canadian laboratory would have to sustain its high level of scientific research and work towards new areas in the future.

FUTURE SYSTEMS

One of Cockcroft's legacies to Lewis was the Future Systems Group – a committee whose title described its function. From the end of 1944 until June of 1946, Cockcroft held regular meetings of this group to discuss possible future projects. In November 1947 Lewis decided to revive the group to discuss possible designs for a new pile.

There were good reasons for Lewis to consider construction of another pile. In October there had been an accident involving NRX. Although the pile itself was not seriously damaged, a great deal of heavy water was contaminated and the pile was shut down for six weeks. Such an accident highlighted the fact that work at Chalk River revolved around a single piece of equipment; perhaps it was time to build another. Added to concern about pile accidents was the belief that the lifetime of NRX would probably be only five years.[54] To design and build a new reactor would take at least that long, so planning would have to start immediately.

Guidance for building the new reactor would come from Lewis and his Future Systems Group. This group of scientists and engineers varied considerably in size over the years. Its earliest meetings included many scientists from the British project including Egon Bretscher, Robert Spence, and Bruno Pontecorvo. The main Canadian scientists were Hugh Carmichael, Hank Clayton, Don Hurst, and Bern Sargent. Later, George Laurence would join the group. These men were all important scientists who would play central roles in reactor development at Chalk River.

Lewis firmly believed that "the initiative for making suggestions for future plans, lies with the staff at Chalk River."[55] This belief, that the scientists performing the research were better equipped to make decisions about Chalk River's future than their political superiors, stemmed from his time at TRE during the war and was not necessarily shared by his boss, C.J. Mackenzie. Although this method of action had worked well in wartime when funds were virtually unlimited, Lewis would later find that more political methods had to be employed to secure funding under normal circumstances. Perhaps indicative of a somewhat cavalier approach to funding was Lewis's advice that, although it was desirable to keep proposed plans within the current annual budget of $5.5 million, "the possibility of a new major project involving additional expenditures should be entertained if there were a strong case for it."[56]

At this first meeting Lewis suggested a number of different types of piles for consideration, ranging from a small homogeneous "slurry" pile using heavy water to one similar to the Harwell pile BEPO – a natural-uranium graphite-moderated pile. Lewis ruled out the Harwell pile since it would be expensive to build and would not teach Chalk River scientists anything new. Even if "the scale of the Canadian project is such that we cannot expect to lead in the development of atomic energy," Lewis commented, any project undertaken should nevertheless keep the staff in touch with the latest techniques. The slurry pile might be considered as a project that could be quickly put into place if NRX broke down, but Lewis's personal favourite was a high-temperature pile using beryllia as moderator.[57]

One difficulty in selecting a reactor was the large variety of possible reactor types. Would a reaction using fast or slow neutrons be better? If the latter, should natural uranium or enriched uranium be used as fuel? Graphite, heavy water, or a gas as moderator? Another choice had to be made about the coolant. Different combinations of these components produced distinct reactor types.

In order to make these decisions, the reason for constructing the new reactor had to be clear. Would it be used largely for pure research, production of fissile material, or production of power? Lewis regarded the proposed reactor mainly as a means for keeping "the Canadian staff in touch with the problems of atomic power" – a rather vague and unhelpful comment. But he added that although Canada at that time had enough water power, he felt they should be prepared for the future when nuclear power might be an important source of energy. It is interesting that Lewis was intrigued with the concept of generating electricity from nuclear power at this early date but realized that the Canadian project could not yet devote itself to that goal.

At this first meeting the central reason for building the new reactor was not stated, although it was clearly crucial in deciding the choice of reactor type. In fact, little was resolved on technical matters; administratively, members decided to break up into smaller groups, which would then report back to the larger committee. Minutes from subsequent meetings of the Future Systems Committee reveal conflicting and changing views on the purpose of the proposed future reactor. At a meeting in January 1948, only two months later, Lewis declared that the committee's objective was to "[decide] on a future reactor which should be in the main line of advance towards the development of useful power and not simply a copy of any existing pile." Les Cook did not agree, arguing that the production of power was too large a project for Chalk River while construction of a high-neutron-flux experimental reactor was not. Lewis answered that outside help might be available for construction of a power reactor but not for a research reactor. Others concurred that a high-flux pile was desirable, however, because of the experimental research that could be performed.[58]

Members of the committee were asked to prepare reports on specific reactor types. Interestingly, although the brief descriptions indicated that natural uranium would be used, heavy water was not at this point proposed as a moderator. But by the end of the next meeting in April, physicist Don Hurst suggested that an improved, high-power heavy-water reactor should also be studied since it would be a good experimental tool, it would produce some U-233, and it could be built at reasonable cost.[59]

At the end of May 1948, C.J. Mackenzie attended a special meeting of the Future Systems Committee. Mackenzie was there as president of the Atomic Energy Control Board, and other board members were also on hand. The AECB was created in 1946 with a mandate to control and supervise "the development, application and use of atomic energy in Canada." Since the National Research Council was operating Chalk River for the AECB, members would have to be convinced of the need for a new pile and the validity of the proposal put forward.[60]

Lewis outlined the case for a new reactor, stressing the limited lifetime of NRX. Noting the high neutron flux and excellent experimental facilities of NRX, he argued that the new reactor should meet these standards and at the same time be a step towards the development of atomic power. Five different reactor types were outlined for consideration but Lewis made it clear that, at this time, the committee backed "a heavy water pile, aiming at higher neutron flux, good experimental facilities, high economy of neutrons and possibly some use made of the power produced." Perhaps most importantly, design and construction of this reactor would make use of the experience gained through NRX.[61]

The response of board members to Lewis's enthusiastic presentation was not minuted and it appears that no decision on the reactor was made at this time. But of particular interest throughout this early planning stage is Lewis's continued emphasis on working towards a power-producing reactor. Although he would remain involved with the Chalk River research reactors, Lewis increasingly was drawn to the possibility of building a reactor to generate electricity.

An early estimate of the time and money involved in the construction of the new pile was made in late 1948 by Ken Tupper, director of the engineering division. Claiming that ninety-five percent of the research carried on at the plant would be halted if NRX broke down, he regarded it as imperative that a new reactor should be built. Tupper estimated that a heavy-water-moderated reactor with high flux would take two full years to design and a further three or four to build. The price tag would be $24 million with another $4 million for heavy water, but he noted that his estimate was crude and "might easily involve an error of a factor of two in either direction."[62]

In February 1949, Lewis outlined additional reasons for building another reactor at Chalk River. For the first time, the larger question of national interest entered into his arguments. He declared it to be in Canada's political and scientific best interest for Chalk River to remain at the forefront of atomic energy development. But to do so, one of the objectives of reactor design must be that it stay within the financial and industrial capacity of Canada. Financial capacity, however, depended upon the value of the fissile material produced; in fact, Lewis viewed Canada's future role as one of supplier of crucial fissile materials to other countries. Although Canada itself would not build any atomic weapons, Lewis does not appear to have had any qualms about supplying the means necessary for other countries to do so. Finally, acquisition of another reactor would let Canada play a part in future developments including war, economic power production, and nuclear energy from light elements.[63]

A number of different reactor designs were listed as still being under consideration, but Lewis stressed that the reactor must be a plutonium producer and provide research facilities for in-depth investigations into the possibility of developing a breeder reactor. This was not the first reference to the future construction of a breeder reactor. This particular type of reactor held a great deal of appeal since it would produce more fissile material than it consumed. In the late 1940s, it was assumed that the uranium shortage would continue indefinitely, making the idea of a breeder reactor a very attractive proposition.

By the next meeting, the three objectives of the new reactor were bluntly stated as high flux, plutonium production, and breeding. To achieve these objectives, four possible programs were suggested. The first was simply to build a reactor to meet all these requirements. This reactor was called NRU – U for universal. Another idea was to have an improved NRX – or NRX' (prime) – which would replace NRX and continue plutonium production, along with a small research reactor (NRF), which would be enriched with Pu and have a high flux. The third choice was a production reactor – NRP – which would be of greater size and power than NRX but have a low flux. Finally, an isotope-separation plant could be built, which would lead to a high-flux production reactor.[64]

In March 1949 the committee presented a definite choice of reactor to C.J. Mackenzie. It would be a heavy-water-moderated reactor with a high flux. This latter requirement meant that the reactor had to be extremely economical with neutrons or use enriched material. Since enriched fissile material was difficult to obtain, components of the reactor could not absorb many neutrons. So, for example, heavy water would be used as both moderator and coolant with zirconium (a low neutron-absorbing metal) for sheathing.

Of central importance to the discussion was the question of how much the proposed reactor would cost. This question was linked to how much money could be made subsequently from plutonium sales to the United States. Lewis was certain that these sales would bring a great deal of money to the project, citing as the value of plutonium anywhere from $200 to $1,000 per gram. It was even possible, he claimed, that a profit might be made, but the costs of chemical processing had to be considered first.[65]

The question of chemical processing had largely been ignored during these discussions. Although mentioned at a number of early meetings, the problems surrounding the development of a chemical-extraction plant to recover plutonium from the irradiated fuel and the costs involved had not subsequently been raised. At a later meeting, it was agreed that the process should not be chosen until experience could be gained from American and British plants. The decision to send the irradiated fuel to the US for processing would eventually eliminate the problem, but it is curious that it did not figure more prominently at this crucial decision-making stage.

By June 1949 the committee was able to outline in greater detail the type of reactor planned. It would be a heavy-water-cooled, heavy-water-moderated natural-uranium reactor with a power output of two hundred megawatts (thermal) (MWt).[66] Although the size of the reactor would be roughly the same as that of NRX, it would differ in a number of respects: it would have a closed cooling system, operate at a higher temperature, and, most importantly, have an on-power refuelling mechanism, which meant that fuel rods could be replaced without interrupting operation of the reactor.[67]

In his sales pitch to the AECB Lewis was particularly careful, once again, to stress the revenue the new reactor would bring in through plutonium sales. It had become apparent in the last year, Lewis told the board, that the estimates of the total irradiation possible had increased from the original of 300 megawatt-days per ton to 1,200 megawatt-days per ton. This would greatly reduce the cost of plutonium production. In fact he suggested the pile might bring in as much as $24 million per annum in plutonium sales. One of his listeners pointed out that these calculations assumed a market for plutonium. Lewis responded by noting that the fundamental market value of plutonium could be taken for granted since if it was not needed for bombs, then it could be used for fast fission breeding of plutonium for the production of electricity. In the end, Mackenzie also agreed that a market for plutonium was virtually guaranteed.[68]

A later report by Lewis indicates that Cockcroft was puzzled by the Canadian decision to build another heavy water reactor. Lewis explained that the British were more interested in power production and breeding than in experimental research and did not therefore need the same high-flux reactor that Chalk River did. For the time being, Lewis was willing to lie low on the power-production front; as Mackenzie told the group, "it would be politically unwise to base the case for the new reactor on power production." Instead, they would stress the view that fundamental research was needed to maintain their position in the future, with the added bonus that the plutonium was marketable.[69]

Work on the new reactor did not advance appreciably in the next year. It had not yet been officially approved and there were simply not enough people available to work on the problems. Staffing deficits plagued the project. Although new graduates were available in May 1950, more experienced people were needed to provide leadership. But the National Research Council salary levels made it difficult to attract scientists. It was clear by early 1950 that the large increases hoped for in both staff and capital expenditures for the future system would not be forthcoming. This meant that more time could be spent on design, but it also delayed the project.[70]

The main reason for delay was the lengthy negotiations with the United States Atomic Energy Commission for its purchase of the irradiated fuel. In a letter to Carroll Wilson, general manager of the USAEC, Mackenzie argued that it made both strategic and economic sense for the US to buy Canada's excess plutonium. As the two countries were "traditional partners in the defences of the continent," and since only the US produced nuclear weapons, it would be safer to disperse production. From an economic viewpoint, it would be better for both countries if Canada sold a more finished product. Finally, this sale would allow Canada to go ahead with construction of a new pile, thus building up a reserve of scientific and engineering manpower of benefit to both countries.[71]

By December 1950 negotiations with the USAEC had been completed and prices agreed upon for Canada's sale of plutonium to, and purchase of uranium and heavy water from, the United States. In the meantime discussions had been ongoing at Chalk River. In September Lewis argued that the reactor program should be speeded up, with completion slated for three rather than four years hence, since the plutonium was needed as quickly as possible. The urgency of the project meant that reactor design would be given priority over current work. In particular, he hoped they could specify the size of the future reactor within eight months.[72]

By early January 1951, cabinet announced its approval for construction of the new reactor. The work was contracted by the NRC to the C.D. Howe Company – a choice that caused some concern. Although Howe had cut his affiliation with the company when he entered politics in 1935, his close ties to Mackenzie invited accusations of patronage. This risk was taken; the company was engaged to perform the necessary engineering services for the new reactor. Meanwhile, Lewis was given responsibility for the scientific and engineering aspects of the future plant. Under him work was delegated to different divisions.

The initial target date for start-up was 1 July 1954, but it was clear by the summer of 1952 that work was already falling behind schedule. The project was an enormous challenge for the C.D. Howe Company, which had no previous experience in building nuclear reactors. This meant that there had to be a great deal of interaction between staff at Chalk River and the company engineers. The main area of disagreement between these two groups occurred over the point at which the final design for particular parts of the plant should be frozen. This was further complicated by the fact that the designs of the more important parts were highly interdependent and so had to reach final stages at the same time. Delays in delivery of key parts, such as heat exchangers, further retarded progress. Attempts by Howe company officials in December 1952 to speed the process by ending alternative studies on certain designs were criticized by Lewis, who felt that design studies had not advanced enough to be frozen. Others were more flexible. Ian MacKay, head of the plant design branch, suggested that only certain specific portions of the NRU design should remain unfrozen. As the debate continued the start-up date for NRU receded further into the future.[73]

Construction of NRU dragged on throughout 1953 into 1954. In January of that year, J.L. Gray, vice-president in charge of administration and operations at the plant, wrote a summary of the problems NRU had encountered in the last four years. The reactor, he announced, would cost fifty percent more than had been originally estimated in September 1950 and the completion date would be at least two years later than the original estimate of July 1954. A great deal of the blame for the growing costs and delays lay with the design engineers of the C.D Howe Company. Their reaction to the first cost estimates made by Chalk River scientists had been to state that it would be "virtually impossible" to spend all the money. But because they were, in Gray's view, unable to recognize the complexity of the design, they tried to solve the problems through mathematical design rather than by "rough design and experimentation." Despite the delays and cost overruns, however, Gray remained enthusiastic about the project.[74]

The novelty of the work, frequent changes in the design, problems with delivery of specific parts – all of these factors contributed to the length of time it took to complete NRU. It was not until 3 November

1957 that the new reactor finally went critical. It had been ten years since the idea was first discussed at a Future Systems Committee meeting.

The NRU reactor was Lewis's first major initiative at Chalk River. He was instrumental in winning approval for the project and determining the type of reactor to be built. Lewis convinced Mackenzie to use the sale of plutonium from the reactor as a means of financing its construction. As chairman of the Future Systems Group, Lewis played a vital part in examining a large number of different reactor types. His capacity for quickly understanding new ideas and making technically sound analyses was essential in deciding on the best reactor type. And Lewis did not hesitate to exert his authority when necessary. Once he had decided that Chalk River needed a new reactor, he focused on that goal until it had been achieved.

Even as Lewis maintained an active involvement with NRU, he was beginning to direct his thoughts towards the future. Since his arrival at Chalk River, he had nurtured an interest in the possibility of building a power reactor. Throughout the early planning stages of NRU, he had emphasized that the new research reactor should be seen, in part, as a step towards power production. Lewis recognized that designing a power reactor was a mission that would enable Chalk River to expand and grow. The research reactors would allow an active program of pure research to continue. But they would also serve as a vital foundation for a power reactor program, without which support of the reactors might become more and more difficult to justify. Lewis remained interested and involved in pure research at Chalk River but, increasingly, he would focus his attention on the development of a nuclear power reactor.

CHAPTER FOUR

Nuclear and Other Pursuits

The town Lewis settled in on his arrival in Canada in September 1946 was still taking shape out of the wilderness. Wartime safety and security concerns had led to the selection of Chalk River, Ontario, as the isolated location of the new laboratory. Most of the scientists and engineers working at the project were housed in a new community ten kilometres up the Ottawa River from the plant. There, at a spot where the river widens, the town of Deep River began to grow.

The town developed rapidly. Clearing of land and construction of homes and buildings began in 1945; by 1948 Deep River already boasted a population of 1,800 and four hundred dwellings "laid about a central shopping centre, near which [were] the Staff Hotel, community hall, school, hospital and dormitories."[1] Roads already built were "wide and gravelled." As new streets were planned and constructed some were given names reflecting the occupation of the town's inhabitants. Thomson Street, Kelvin Crescent, and Faraday Avenue can today be found in Deep River.

Deep River was planned to be a model town, a perfect suburban community carved out of the wilderness. Its isolated location had certain disadvantages in the early years, but as the town grew it acquired many of the conveniences of city life. By 1948 the community centre boasted a library, coffee shop, bowling alleys, and a hall for movies. For many it was the ideal life. The plant was only a short drive away. Weekends and holidays could be spent sailing or swimming in the river, walking or skiing in the woods, or simply enjoying being away from the big city. Perhaps there were too many blackflies in May and too many very cold days during the winter, but for those who enjoyed the outdoors, it was a good life.

While married couples vied for the newly built houses and contributed to the record-high birth rate in the town, the unmarried scientists and other employees lived at the staff hotel. Lewis joined their ranks when he arrived in the fall of 1946. Life at the hotel was relaxed and communal. Meals, served in a large dining area, provided an opportunity for people working in different branches to meet. On Sunday evenings, residents gathered in the lounge and listened to records. Lewis joined them, bringing letters to write while enjoying the classical music.[2]

As he had when he was younger, Lewis struck many people as aloof, even cold. He was a reserved man, but much of his shy demeanour stemmed from his inability to make light conversation in social settings. He did make an effort at socializing but was not entirely successful. One colleague recalled an incident at the staff hotel when a brave young lady approached Lewis at a party and sat on his knee. Lewis, deeply involved in a scientific conversation with a colleague, simply ignored her and continued his discussion.[3]

Although Lewis remained a bachelor, he moved after a few years from the staff hotel to a two-storey white clapboard house on Beach Avenue. Running parallel to the river, this street housed the senior scientists at the plant. Its large homes with gardens stretching down to the river's edge made it a desirable area to live. Lewis asked his mother (his father had died during the war) to join him in Deep River. Mrs Lewis adapted quickly to life in town and looked after the more practical aspects of Lewis's life, leaving him free to concentrate on his work.

Lewis devoted most of his time and energy to work but he did have a few hobbies. In his Cambridge days, he was known as a fast-car enthusiast and owned a Hudson. He was one of the first people in Chalk River to own a car, which he used instead of the bus to get to work. In the early days Lewis skied, bowled, and curled during the winter; in the summer he tried his hand at sailing.

Sailing was a popular social activity among the scientists in town. Most sailed small, home-built boats called Y-flyers. Lewis joined the weekly races in his boat, named *Celia* (the name gave rise to much speculation but no answers), often with Hank Clayton as crew. Remembered as a taciturn and not very good sailor, Lewis nevertheless participated doggedly in these competitions. Later he would lend his boat to students in return for their help in keeping *Celia* shipshape.[4]

Lewis was an avid hiker and spent most of his holidays walking in Canada's national parks (Revelstoke was a particular favourite) or in England's Lake District. But he was definitely more a man of the mind than the body. His main interest apart from his work at Chalk River was the Deep River Library. A voracious reader himself, Lewis thought it very important that the new town should have a good library. In May 1947, shortly after he arrived, a meeting was called to elect six people to the board of the new Deep River Library. With Lewis as chairman, the library was registered as an association library under the Ontario Library Act.[5] One fellow board member recalls that, as chairman of the library board for over three decades, Lewis was "the moving spirit of the library."[6] Board meetings were run in the orderly and decisive manner characteristic of him, and their deliberations bore the mark of his strong personality. For example, Lewis was very assertive during the process of choosing books – a trait that occasionally led to charges of censorship. In particular, he objected to buying books about nuclear power if they were written by scientists he did not respect. And yet if other members felt strongly enough about a given book, he would back down. His control as chairman was not complete but his personality and position in the town were such that it was difficult to stand up to him. The fact that meetings were usually held at Lewis's home with his mother serving cucumber sandwiches and cake to the assembled members tended to reinforce his position.[7]

Lewis also carried the day in the library's periodic reviews of its status within the Ontario library system. When pressured in the mid-1950s to become a "public" rather than an "association" library (a change entailing new funding arrangements and reduced independence for the board), Lewis's preference for remaining an association prevailed.[8] The Ontario Public Libraries Act of 1966 reopened the issue, now framed as a choice between becoming a "public" library or turning private. After carefully weighing the pros and cons, Lewis argued in favour of becoming a public library – a decision that the board adopted.[9]

Despite, or perhaps because of, Lewis's strong control of the board, the Deep River Library flourished. Its membership grew steadily and more books circulated each year. By April 1954 an extension was necessary;[10] by 1971 it had outgrown the community centre altogether and the kitchen and dining-room area of the old staff hotel was converted into a new library building.

Lewis had other ideas for expansion that went beyond the town limits. In his view a basic function of any library was to make books, and therefore information, accessible to the public. In large cities this was not a problem, but for people living in remote regions of the province books were difficult to acquire. Lewis copied the idea from the Barry's Bay Public Library, which had set up a reading room in Wilno, a nearby small town. Largely on Lewis's encouragement, the Deep River Public Library set about establishing reading rooms from

Chalk River to Deux Rivières, a distance of eighty kilometres further along Highway 17. Each reading room would be equipped with various reference materials and would be loaned about two hundred books, which would be changed periodically. Local residents greeted the reading rooms enthusiastically and it was hoped that they would later become libraries.[11]

Lewis vigorously promoted reading rooms in the last years before his retirement and remained actively involved on the library board until 1983. Some people undoubtedly resented his constant presence as chairman of the library board but if so, they retired. Those that remained remember him as a decisive chairman who helped the library become one of the best small-town libraries in Ontario. To honour his contributions, the library was renamed the W.B. Lewis Public Library during its fortieth-anniversary celebrations.

Lewis enjoyed these town activities but continued to devote most of his attention to developments at the plant. As one colleague commented, "his life was his work."[12] This commitment was linked to his religious beliefs. It has been suggested that Lewis's sense of religious duty helped to guide him towards working on the development of nuclear power. Having witnessed the horrible destruction that could be wrought by harnessing the power of the atom, Lewis hoped at Chalk River to create constructive uses for this same energy.[13] As will be seen, this belief also seemed to lie behind his argument for supplying nuclear power to underdeveloped countries.

Lewis participated fully in church activities in Deep River. Until 1962, he and his fellow Anglicans had shared a building with the United Church congregation. In 1959, Anglicans in Deep River formed a committee to discuss building a church of their own. Although Lewis was unable to attend the meeting at which final arrangements were outlined, he sent a card to another member stating, with his usual thoroughness, his opinion on the questions under discussion – complete with biblical references.[14] Lewis's mother laid the cornerstone for St Barnabas, the new church, in 1962, and both remained deeply involved in its activities.

The common thread running through Lewis's involvement in the library, his religious devotion, and his scientific endeavours was his ability to focus solely on the task at hand. His colleagues still recall his long hours at the plant and disciplined work habits. On a typical day he would remain at his office until five in the afternoon, return home for supper, put on a classical record, and sit down to work all evening. On business trips he would use the time spent in waiting rooms and aboard airplanes as opportunities to question colleagues accompanying him about the results of their work. Other scientists recalled with frustration that if Lewis did not understand a particular subject, he would borrow the relevant texts and spend the weekend mastering it.[15]

Lewis's capacity to understand new material was much better than his ability to communicate information to others. Swaying from one foot to the other, he delivered his speeches in a monotone and in a manner that often confused his audience. He was not at ease giving lectures and instead enjoyed heated discussions during committee meetings. Although ruthless when grilling staff for information and attacking their work if he thought it to be flawed, Lewis believed that these intense discussions, perhaps reminiscent of wartime Sunday Soviets, were the most effective method for finding answers quickly. He expected his staff members to disagree with him, too, and in the early years would turn to them and exclaim, "Will no one argue with me?"[16]

Shortly after his arrival, Lewis established a series of director's reports, lectures, and memoranda to supplement discussions with his colleagues. During his career Lewis would write over three hundred of these articles for circulation throughout the plant and sometimes for publication in scientific journals. Many of the documents were highly technical in nature and likely aimed at senior staff members at Chalk River and close collaborators like Cockcroft. It is possible that their most useful function was to help Lewis, tireless worker that he was, to clarify ideas by putting them on paper. Lewis continued to write these articles throughout his career at Chalk River and established another series of them during his retirement.

While Lewis sorted out his ideas through his document series, the scientists and engineers he worked with defended their ideas in lively discussions. Many disliked Lewis's confrontational style of management but it did tend to ensure that subject areas were well understood. This expertise would be essential in tackling problems in what would soon become the central focus of Chalk River: the development of a nuclear power reactor for the economic generation of electricity.

BREEDER REACTORS

In August 1945, the explosion of two atomic bombs in Japan demonstrated to the world the enormous destructive power of atomic energy. Initially horrified by its potential for devastation, people were soon swept up by the promise of atomic energy. Newspaper and magazine articles discussed plans for atomic-powered cars and airplanes, and schemes to change the climate and melt the polar ice cap were seriously considered. One magazine predicted that soon bulky hydroelectric plants would be replaced by "small, neat energy-producing buildings." Another described a small household atomic energy unit that would heat and cool a home for a decade. There seemed no limit to the wonders of this scientific discovery.[17]

Exaggerated predictions of the benefits of atomic energy would soon fade but certain hopes remained. Most durable was the belief that cheap and abundant power could be obtained from the atom. Lewis was fascinated by this possibility. He had recognized the power potential of nuclear fission while still chief superintendent of the Telecommunications Research Establishment, but he believed that it would be at least five years before fission research would bring about a net economic gain. For this reason, he had argued, Britain ought to devote itself to more immediate problems. TRE's role would be to provide the electronic control and experimental equipment needed at atomic power plants. With TRE's future uncertain in the postwar years, Lewis hoped to link it with the proposed atomic energy establishment in the electronics area. But during his brief tenure as chief superintendent, Lewis had little time to follow up these possibilities.[18]

Lewis's involvement with atomic energy research changed dramatically once he was appointed Chalk River's new director in September 1946. By October he had written a report outlining possibilities for future projects. In this early report Lewis was already marshalling the points in favour of constructing another reactor at Chalk River. But the question remained as to the type of reactor to replace NRX. Of the many ideas put forward, Lewis focused on those he saw as steps towards the production of economic nuclear power.[19]

Lewis was not alone in pondering the possibilities of producing electricity from a nuclear reaction. The idea that the energy produced by splitting the atom could be harnessed to generate electricity had been considered during the war. Wartime pressures had forced scientists to work exclusively on building a bomb, but by the fall of 1946 scientists in Great Britain, the United States, France, and the Soviet Union were studying the many different reactor types possible. As Margaret Gowing has noted, the long-term problem of nuclear power was how to provide technological solutions to problems economically; in order to be worth developing, nuclear power had to be competitive with coal and oil.[20] A further difficulty was the enormous choice of reactor types – approximately one hundred different kinds. American, British, French, and Russian efforts were further complicated by the need to coordinate research with their weapons programs. But despite differences in scale and objectives in the realm of nuclear research, scientists all shared an interest in the breeder reactor.

The breeder reactor was so named because it would produce more fissile material than it consumed. For example, a fissile core of plutonium surrounded by a "blanket" of thorium could, in principle, produce the fissile isotope U-233, by neutron capture in the thorium, at a faster rate than the rate of destruction of plutonium in the core. A central problem with the breeder was the expense involved in chemical extraction and fuel reprocessing, which cast doubt on its economic viability. But because of a perceived shortage of uranium in the early postwar years, scientists of all countries took the breeder very seriously.

A report that Lewis brought over from England offered guidance and served as a common point of departure for the Canadian and British programs. Prepared in September 1946 by C.H. Secord, an official at the Ministry of Fuel and Power, the report outlined the possibilities and problems of generating electricity by using nuclear power. It argued in favour of a breeder reactor both because of the uranium shortage and because a small proportion of the world output of thorium could replace all other sources of energy in Great Britain. A serious initial problem lay in acquiring enough fissile material – plutonium or uranium-235 – to initiate breeding. Secord proposed an optimistic timetable for the replacement of conventional sources of energy by nuclear. Given a rapid breeding rate, he argued that by 1965–70, Great Britain could increase its energy supplies by one-third by replacing existing supplies through the use of nuclear power.[21]

Scientists and engineers from Chalk River discussed the Secord report at a meeting chaired by Lewis in early December 1946. Members of both the Canadian and British teams were present, including Hugh Carmichael, Hank Clayton, Don Hurst, Egon Bretscher, and Bruno Pontecorvo. It was Lewis's view that, despite the report's "obvious shortcomings" and the criticism it had received in Great Britain, it nevertheless appeared to be "highly important" and worthy of serious consideration.[22]

A subsequent meeting clarified the main problems envisaged in power production. Of central importance for all reactor types was the question of heat removal. In order to maximize use of the nuclear reaction it was essential to design the most efficient heat-removal system possible. Problems arose from the lack of knowledge about how materials would behave when subjected to the high neutron flux and high temperatures within the reactor. Further study was also needed on the diffusion of fissile material and fission products in the reactor and on the question of their disposal. Given the fundamental nature of these problems and the need for more basic research, the group questioned the time scale projected in the Secord report and suggested increasing it by ten years. This meant nuclear power would replace conventional power sources by 1975–80. Finally, the Chalk River scientists agreed that the lack of information on reactor costs and on the future price of coal made it difficult to give an opinion about the economic viability of nuclear power. But the meeting closed on an optimistic note, with a general feeling that there would not likely be a great disparity in cost between the two sources.[23]

In a letter to Cockcroft, Lewis reaffirmed his belief that Secord had underestimated the difficulties involved in obtaining power from a nuclear reaction. "It is the difficulty of releasing energy from nuclear fuel, keeping under control the rate of release and the disposal of fission products, which is the key to the economic development and application of nuclear power," he wrote. "Secord's paper assumes that this is too easy."[24] Lewis discussed two possible routes to power development, both of them using breeder reactors. The first involved a fast fission reactor using plutonium, the advantage of which was that plutonium could be, and was already, produced (for weapons purposes) in slow-neutron natural-uranium-fuelled reactors, and this could be used to start the breeding program. Its disadvantage was that fast reactors represent a much more serious engineering challenge than slow-neutron reactors. The second possibility was based on the fact that uranium-233 can breed in slow-neutron reactors (while plutonium cannot), the disadvantage being that building up a stockpile of uranium-233 is less straightforward.

Whatever the route to breeding, neutron conservation is of paramount importance. Unproductive loss of neutrons means a corresponding loss in the amount of fissile material produced, which has to be greater than the amount consumed in the fission process. The net production rate sets the rate of expansion of a power-generating system based on breeder reactors. Lewis was quick to realize the importance of neutron conservation – even for nonbreeder reactors – and it became a central concern and a guiding principle for him. He also emphasized the difficulty for any type of breeder reactor of handling and processing nuclear fuel.[25]

In March 1947, Lewis published his latest views on atomic power reactors in his first director's report, "World Possibilities for the Development and Use of Atomic Power." Citing one American and two British reports, Lewis observed that little had been published on the topic thus far. As C.J. Mackenzie and C.D. Howe had hoped, the

Canadians were in the running at the beginning of this new technological development.

A breeder reactor remained Lewis's central interest. In order to start a breeding process, however, an initial stock of fissile material was necessary, and here politics interfered with the reactor's rapid development. Obviously, Lewis noted, the most efficient way to build up a stock of fissile material was to do so in one place. But although the United States had undoubtedly started a stockpile, its "isolation ... on political grounds" forced other countries to build their own. The United Kingdom had committed itself to this course (and there, as in the United States, the material was for weapons, not breeding), but Canada had not yet decided whether to start its own breeding enterprise.[26] The problem was that this first phase was enormously expensive; Lewis quoted figures in the range of $100 million to $300 million as the initial expenditure; further funds would be necessary to develop and build breeder reactors and "secondary" reactors – a term Lewis gave to reactors consuming prepared fissile material and delivering only power.[27]

How could Canada justify these enormous expenditures? Lewis based his argument upon the very large amounts of energy stored in a small amount of fissile material. In a table, he compared the total energy stored per tonne of nuclear fissile material to that of oil and coal. According to Lewis's calculations, the fissile material had energy nine orders of magnitude greater than either of the others. The drawback was that while the energy from coal and oil was easily released, that from nuclear fuel was very difficult to obtain and involved health precautions and protective structures. But to Lewis, these difficulties were simply a challenge to be met.

In fact, the central difficulty was one that Lewis did not highlight: paying for the reactor. His estimates for the cost of the reactor were of the order of $100 million at a time when the total federal budget was $2.63 billion and the budget of the Department of Reconstruction was only $2 million. It is also unclear how Lewis derived his estimates at a time when there were no precedents to work from. Lewis, it appears, was not overly concerned with these problems. He focused as usual on the scientific challenge and let those above him come up with the money.

Despite the difficulties involved in producing nuclear-generated electricity, Lewis argued that the inevitably increasing world demand for electricity justified the effort. He readily admitted that in certain circumstances nuclear power was not feasible. It could not be used to provide locomotive power or to run small factories that did not use much electricity. But it might displace coal and oil if large installations were built in areas where there was little water power. The heat from a nuclear power plant might also be used instead of electricity in certain production processes requiring heat, or to heat densely populated cities.

Lewis particularly stressed the growing use of electricity in underdeveloped countries. Nuclear energy, he believed, could bring enormous benefits to these countries. Easy transport of the fuel meant that distant countries could be supplied without difficulty. Further speculation led him to consider whether nuclear power might be able to alter extreme climate conditions within a given region. The question was, as Lewis noted, whether society thought this a worthwhile use for nuclear power. Lewis pointed out that climate control was the "upper limit" for what nuclear power might achieve.[28]

Lewis also touched on a problem that would later become very serious and of central importance to the nuclear industry – proliferation. He believed that, while during the last war countries had sought to develop their industrial capacity to its utmost, in future wars the atomic bomb would render superiority in industrial power useless. In fact, as he wrote, "a tolerable defence against [the atomic bomb] cannot be achieved by any foreseeable application of the physical sciences and industrial resources." Because of this, Lewis argued in a roundabout way, it was useless to attempt to stop the spread of knowledge about nuclear power because "the limitation of industrial resources in other countries does not in the long run afford military defence against them, because the fraction of the industrial resources needed to turn over to atomic warfare is only small."[29]

These political problems remained in the future. Of immediate interest was the reactor Lewis suggested for consideration. Lewis proposed a slow neutron "pile of thoria and beryllia seeded with Uranium 233 and cooled by helium," which would produce more U-233 than it would consume. A large stainless steel pressure vessel would contain beryllia to act both as moderator to slow down neutrons and as reflector to limit their escape. Further supplies of U-233 would be bred from thorium. Channels would be "left in the oxide through which helium is forced under pressure as a coolant and working gas for converting the heat to mechanical power." Of crucial importance to the operation of the breeder pile was economy of neutrons. But if it were constructed properly, Lewis hoped that the stock of U-233 would double every four years.[30]

Lewis's first director's report had some lacunae. He was forced, as he had been when estimating the cost of building a reactor, to outline this reactor design with very little past experience. Any work on breeder reactors in other countries was not shared at this early date. The result was that details were sketchy and assumptions had to be made. Nevertheless, Lewis wrote this report six months after his arrival at Chalk River. It is striking that he was able, at this early date, to outline a possible reactor program that could be discussed and criticized by his colleagues.

Canada, through Chalk River, could play an important part in answering some of the many questions involved in mastering this new technology. The NRX pile could produce supplies of plutonium and uranium-233 for early pilot experiments. Equally important, it could be used to determine crucial physical constants necessary for calculations. Lewis foresaw a dual role for Chalk River: scientists would be involved both in planning the design of a power reactor and in performing fundamental research. In this way, Chalk River would be similar to laboratories in the United States and Great Britain.

Lewis supplemented his first report with a second, entitled "The Future Atomic Energy Pile," in September 1947. Although still undecided about the proposed reactor's exact form, Lewis continued to favour a breeder reactor. It would operate in conjunction with a chemical plant that would remove fission products, recover fissile material, and refabricate fuel. Lewis predicted that reprocessing would be required after only a small percentage of the fissile material had been consumed and would necessarily be repeated on a continuous basis. An efficient process would therefore be required, and Lewis suggested simplifying the refabrication step by using oxide or carbide fuel to avoid having to reduce source and fissile material to metal.[31] His report outlined the extensive research effort necessary to surmount the challenges involved in the design of such a reactor. NRX would be a critical asset. Cross-sections of all materials to be used in the future pile had to be accurately determined, and NRX would have to run at full power in order to provide the necessary amounts of plutonium and uranium-233 in the desired time. New chemical processes for the removal of fission products had to be developed and engineering problems of heat removal solved. Finally, the effect of irradiation and high temperatures on all the materials to be used in the pile had to be assessed.

Lewis discussed his second report with all the branch heads. In a covering letter to Mackenzie, Lewis admitted that they felt he was underestimating the magnitude of the program. He suggested that it might be necessary to link the project "with some larger resources such as the United Kingdom Atomic Energy Project," but he hoped that they might also try to attract more "first rate staff."[32]

George Laurence commented in detail upon Lewis's report.

Laurence had recently returned from New York where he acted as special advisor to Canada's United Nations mission. During his absence, Lewis had been appointed to replace Cockcroft as director. As the senior Canadian physicist at the plant, Laurence possibly coveted the position himself and might have resented Lewis. But although Laurence and Lewis were known for their heated exchanges while travelling to and from the plant, the two worked well together.

Laurence agreed that Chalk River should be involved in developing atomic power for peaceful purposes but felt that Lewis's paper should clearly outline a specific design and how long it would take to achieve it. Laurence argued that in the next three years, Chalk River scientists should design a reactor capable both of supplying heat to operate machinery and of replenishing its fuel supply. If after this period it was recognized that the particular reactor concept was not worth pursuing, they should switch to another, building upon the experience gained. At this stage, Laurence appeared to be planning on a smaller scale than Lewis, although they were in agreement that the breeder reactor held the most promise.[33]

The debate surrounding the most promising type of reactor continued at Chalk River. At a meeting of the Future Systems Group in April 1948, people presented their favoured reactor type. Again, although the details of types of fuel, coolant, and moderator differed, the reactors were all breeders. Further information about the different materials used in building a reactor was gleaned from the extensive irradiation program underway at Chalk River. It was vital to know how these materials would react under the high-flux and high-temperature conditions existing within a reactor.[34]

Lewis and the senior scientists at Chalk River appear to have remained committed to the breeder reactor during 1948 and 1949 but little was done to secure approval for building it. Since it was still not certain that NRX's replacement – NRU – would be approved, it was not a propitious moment to introduce a plan for another expensive reactor. It also appears that Mackenzie did not encourage Lewis to devote a great deal of time to the question of building a power reactor. In a letter written in March 1948, he noted that "we in Canada are not particularly interested in atomic power generation at the moment, and the only people who appear to have the power generation as a high priority are the English."[35] But although Lewis was giving priority to the more pressing projects, the prospect of atomic power remained his primary goal. He continued to push the case for building a breeder reactor to generate electricity when a parliamentary committee visited Chalk River in November 1949. He told the parliamentarians that the "gleam in the eye of the atomic scientist" was power comparable in magnitude to that available from coal, oil, and water but available anywhere since the transportation costs of nuclear fuel would be negligible. Lewis continued to sell the idea of a breeder reactor and highlighted its positive attributes but did not give the committee a feel for the enormous problems yet to be overcome.[36] It was, however, a necessary political move; if the Chalk River program were to expand into power reactors, government money would be needed.

Over the next eighteen months Lewis's devotion to the breeder reactor waned, owing to several developments. Much of the early enthusiasm for breeders derived from a widespread belief that there was little uranium to be mined in the Western world. This made thorium a desirable substitute. By the early 1950s, however, the uranium shortage had proved less acute than earlier believed. This made the need for a breeder reactor, and for thorium, less pressing. Further, it is clear from Lewis's reports that the engineering and design of a breeder reactor would have been an extremely difficult project requiring more manpower than was available at Chalk River. In particular, fuel reprocessing was proving to be both complicated and expensive. Of greater significance, however, was the fact that the scientists and engineers at Chalk River had experience and knowledge in building and operating a heavy-water natural-uranium reactor – not breeder reactors.

This final point was the central thrust of a memo George Laurence sent to Lewis in March 1950. Laurence argued that plans for building a breeder reactor should be postponed. Instead, effort should be concentrated upon the proposed new reactor (NRU) which would provide "facilities and opportunities for the development work necessary for design of the first power plant."[37] A serious problem with a breeder reactor was the time required to build up a stock of fissile material to fuel it. Such a supply could be gathered from the new reactor and used later to fuel a breeder.

Laurence believed the best hope for building a power reactor in Canada lay in the heavy-water natural-uranium model. He argued that Canadian scientists should not look to the Americans or the British for direction as they would both use enriched fuels. Furthermore, Laurence was convinced that engineering and material developments at Chalk River would result in more efficient power production. For example, he suggested that use of zirconium for sheathing would bring greater efficiency and that more heat could be extracted per tonne of uranium.[38]

Lewis was strongly influenced by Laurence's arguments, which came at a time when encouraging experimental results at the plant suggested that an economical power reactor might be possible. In a report written in August 1951, Lewis abandoned the idea of a breeder and instead proposed construction of a heavy-water natural-uranium reactor. He based his conversion on three crucial advances that had occurred during the last three months. The first, hinted at in Laurence's memorandum, concerned the amount of heat extracted from the fuel rods. Contrary to expectation, it had been found that a great deal of heat (up to 2,700 thermal megawatt-days per short ton) could be extracted from uranium metal without reprocessing the fuel, which would be a costly and complicated business. Lewis calculated that with this amount of heat available, uranium fuel would be three to four times cheaper than coal or oil. This price would drop further if a value was assigned to the plutonium or depleted uranium in the rods. Secondly, while designing NRU, a method had been developed for changing fuel under power, which meant the reactor could be operated continuously. Finally, experiments had proven it possible to operate a heavy water system at high levels of irradiation, high temperatures, and high pressures.[39]

By means of elaborate calculations, Lewis demonstrated that the eventual cost per kilowatt-hour of electricity generated by his atomic power reactor would be comparable to that of a coal-fired plant. The power reactor was to be a "heavy-water natural uranium reactor pressurized to over 1500 lb / square inch and with the coolant heavy water heated to 550F." It would have an output of four hundred megawatts (thermal). The outer shell of the pressure vessel would be made of steel capable of withstanding high temperatures and pressures. Fuel elements would be sheathed in either aluminum or zirconium. Lewis estimated the capital cost of the reactor, including heavy water (but not uranium), at $20 million, not allowing for interest-rate changes during construction.

In this report Lewis put forward a firm proposal with a set price tag, but he did not claim that all problems had been solved. One difficulty presented by atomic power plants was that they occasionally shut down at unscheduled intervals. This could disrupt an electrical power grid and Lewis admitted the need either to have reserve generators ready or to link the nuclear power reactor with a conventional coal-fired plant. To diminish the costs of the reactor, Lewis suggested that plutonium be extracted from the fuel slugs. Undaunted by these potential drawbacks, Lewis's enthusiasm for the basic reactor design was evident.[40]

A few months later Lewis expanded upon these ideas in another report. He discussed four different approaches for trying to produce economical power from nuclear energy: large scale, plutonium production, mobile reactor, and natural uranium. He argued that the large-scale approach, which was pushed in the early postwar years by Cockcroft, Secord, and himself, was flawed precisely because it relied upon breeding, likely an expensive process. The plutonium-production approach depended upon a reactor that would produce both power and plutonium. For this reactor to be economic, the cost of building a breeder would have to be low and the value of the plutonium produced high. This meant the cost of the power produced was dependent upon the vagaries of plutonium prices. A further difficulty in relying on plutonium extraction was that for weapons-grade plutonium, the fuel would have to be removed after an irradiation time that was five times shorter than would otherwise be possible, so that five times the uranium would be needed to get the same energy. Lewis discounted the "mobile reactor," which he described as a small reactor operating at a high temperature, because its development would not lead to large-scale operation.[41]

In Lewis's mind, the "Chalk River approach," as outlined in his report "An Atomic Power Proposal," was the best route to follow. Even in the few months since the report was published, advances had been made. Experiments had proven it possible to extract even more heat from a given amount of natural uranium. Calculations indicated that a pound of uranium gave four times the heat of a tonne of coal but cost the same amount. Problems remained, however. The three conditions specified for the Chalk River approach – continuous operation during fuel changes, irradiation to three thousand megawatt-days per tonne, and operation at 550F in a high flux – had not been achieved simultaneously. While Lewis felt that some of the problems only needed further engineering to be worked out, he worried about the metals involved standing up to the high temperatures and fluxes. But the facilities at Chalk River were well equipped to provide answers to these questions.[42]

These two reports marked an important change in Lewis's thinking on future power reactors. His interest in the possibility of generating electricity by use of a nuclear reactor dated to his early days as director of Chalk River. Lewis, like scientists in many other countries, had looked to a breeder reactor. But the disadvantages of a breeder – its complicated design, the expense of reprocessing fuel elements, and the need for a core of fissile material to start – had become increasingly apparent. Further, designing a breeder would not fully exploit the knowledge and experience gained from building and operating the NRX reactor.

In light of this new situation, and prompted by Laurence, Lewis prepared his reports. Gone was the breeder reactor; in its place was a natural-uranium heavy-water reactor. It was this type, Lewis now argued, that Chalk River scientists should develop into an atomic power reactor.

In retrospect, these reports signalled a significant turning point. For the first time, the two central material components of future CANDU power reactors – natural uranium and heavy water – were in place. Lewis must be credited with bringing together a number of different developments at the plant to form the basis of these reports. Nevertheless, it is clear that for Lewis, this power reactor design was not an inevitable outgrowth of the research reactors. Like most other scientists at the time, Lewis's first allegiance was to the breeder reactor. It seems likely that George Laurence helped to guide Lewis back to reconsidering natural uranium and heavy water as elements of a power reactor. Experimental results then convinced Lewis that this was the best route to power. Once this connection was made, he stuck with it tenaciously.

The reports did more, however, than introduce Lewis's new preferred reactor type. They outlined a mission for the project. Chalk River had been established during the war with a specific purpose in mind: to build a plutonium-producing reactor. With the end of the war, the urgency of this mission had faded. Instead, scientists at Chalk River reaped the advantages of having a world-class reactor for experimental use. Basic research would remain an important part of Chalk River's program; the decision to build NRU was proof of that. But if Lewis had his way, Chalk River would soon set its sights on a new and challenging goal: a heavy-water natural-uranium nuclear power reactor capable of generating electricity at competitive rates.

CHAPTER FIVE

"The Gleam in the Eye of the Atomic Scientist": Towards a Power Reactor

By 1951 the character of the Chalk River project had changed considerably. The approval that year by the board of directors to build another reactor – NRU – ensured that Chalk River would have a future stretching beyond the lifetime of the NRX reactor. Chalk River had proven itself to be an excellent nuclear research establishment. It was also growing in size; on 1 April 1952 there were over 1,300 people employed at the project. The new staff needed for NRU would raise this number to over 3,000, which would give Chalk River more employees than all the other laboratories of the National Research Council combined.

These two factors – the growing size of the atomic energy division and its more certain future – triggered a decision to separate the division from the NRC and form a new Crown company, Atomic Energy of Canada Limited (AECL). C.D. Howe told Mackenzie, at the time president of the NRC and leader of the atomic project, that he must choose between the two and suggested that he head up the new Crown company. Since Mackenzie was already president of the AECB, this would consolidate atomic energy matters under one man. Mackenzie agreed and on 1 April 1952, Atomic Energy of Canada Limited was formed.[1]

The new company was divided into three sections: Research and Development, Administration and Operations, and Medical Services. Lewis became vice-president, Research and Development and by 1954 he was formally joined by J. Lorne Gray, who was made vice-president in charge of Administration and Operations. Gray had come to Chalk River in 1949 after working for Mackenzie at the NRC for two years. Initially, he was put in charge of administration but in 1954 he also took on operations, which had been under Lewis's control. Lewis was apparently content with the new arrangement as it allowed him to focus his energies on research, leaving administrative and political problems to Gray.

Lewis and Gray worked together as a team until Lewis retired in 1973. Their partnership was successful because each possessed talents the other lacked. Lewis, the scientific master, had a detailed knowledge of technical developments at the plant and clear ideas on its future direction. Gray shared Lewis's ambitions for Chalk River and had the political skills necessary to achieve them. Furthermore, Gray had a keen economic sense. Lewis's championship of projects rarely included a clear understanding of their cost and whether funds were available. Gray understood that science was expensive. He was able to rein in Lewis's enthusiasms and help guide projects successfully through the political maze. To a certain extent, Lewis recognized both his own deficiencies and Gray's talents. When Bill Bennett, who had succeeded Mackenzie as president in 1953, resigned his position in 1958, Lewis was content to let Gray take over. Gray, he realized, could better deal with the politicians, leaving him free to concentrate on science and engineering.

Lewis had a different title within the new company but his job remained the same. Research would remain of fundamental importance to AECL. Added to this, however, was the growing interest in the development of a nuclear power reactor. Both these avenues of research at the plant depended heavily on the NRX reactor. In 1952, the fragility of this situation was plainly demonstrated.

THE NRX ACCIDENT

Only six months after the formation of the new Crown company, an event occurred that threatened the future of the entire Chalk River project. On 12 December 1952 an accident in the NRX reactor left it completely disabled. Initially it was not clear that it could be salvaged and with the new reactor – NRU – still in the design stage, the project was totally reliant upon the now-crippled reactor. Lewis reacted quickly in the emergency and decided that the reactor would be rebuilt. His leadership during this difficult process has since been praised as a highpoint of his career.

The reactor was operating at low power when the accident occurred. Experimental measurements were being recorded to compare the reactivity of long-irradiated fuel rods with that of a fresh rod. The accident itself can be blamed on a combination of mechanical and operating errors. Problems started when an operator unintentionally raised a number of shut-off rods. These rods are made of a substance that absorbs all the neutrons that strike it. When inserted in the reactor, their presence rapidly diminishes the number of neutrons present and stops the chain reaction. This error – the raising of the rods – was not, in itself, enough to cause an accident, provided that it was corrected fairly promptly. But when the supervisor tried to remedy the error, the shut-off rods did not all drop back into position. The situation was compounded by another mistake that resulted in another bank of shut-off rods being raised, sending the reactor far over critical. The assistant supervisor realized this and tried to drop the shut-off rods back into the reactor, but they did not return as rapidly as necessary. In the control room, everybody realized they had a crisis on their hands and it was decided to dump the moderator. This finally stopped the chain reaction but not soon enough to prevent major damage to the reactor.[2]

The reactor was a shambles. The high temperatures reached when power was at its peak had melted the aluminum sheathing around some of the uranium rods. Another rod had blown itself apart, leaving highly radioactive uranium scattered around. Highly radioactive cooling water was pouring out of the reactor vessel at hundreds of gallons per minute and accumulating in the basement.

Lewis reacted well in this crisis situation. Although not on site at the time of the accident, he quickly took control and decided that the reactor would be salvaged. Within ten days a pipeline a mile and a quarter long was built to the disposal area and the roughly one million gallons of radioactive water in the basement pumped out. Canadian and American military personnel were brought in to help with the clean up. The damaged rods and radioactive calandria had to be removed. They were so highly contaminated that each worker could only remain in their vicinity for a minute before being replaced.[3]

Lewis's decision to save the reactor was, in retrospect, the right one. The NRX accident was the first major reactor accident of the nuclear era. There was no experience to guide the decision. One colleague, P.R. Tunnicliffe, recalled working with Lewis trying to interpret the accident records. Lewis, Tunnicliffe recalled, remained interested on a day-to-day basis in developments concerning the reactor.[4] He was the driving force behind the reconstruction of NRX. George Laurence also praised him for the "promptness and decisive and uncompromising leadership with which he set the staff to work to clean up the radioactive contamination." In Laurence's opinion it was this rapid action that "helped to win continuing government support for the nuclear program."[5]

TOWARDS A POWER REACTOR

In 1951, Lewis's report DR-18 suggested a possible model for a natural-uranium heavy-water power reactor, an idea he developed further in subsequent papers. Through these reports and his speeches to interested groups, Lewis tried to convince everybody of Canada's need for power reactors. He received encouragement from a British colleague who had studied his reports. Christopher Hinton, the engineer in charge of designing, constructing, and operating the plants to produce fissile materials in Great Britain, wrote that he believed the "long term solution to the problem of producing power from atomic energy lies in the development of fast fission reactors."[6] Yet he agreed that the fastest way of producing power from atomic energy was from thermal reactors (i.e., reactors using thermal instead of fast neutrons). He praised Lewis's reports and told him that if enough heavy water was available, then "the best possible thermal reactor is the one that you are proposing."[7] It should be possible, Hinton believed, to generate power in this way within five years.

During the summer of 1952 the Future Systems Group met to decide on a course of action. As Lewis explained at the beginning of the meeting, the larger group would divide into a Nuclear Physics and Reactor Engineering group and a Metallurgy and Chemical Processing group. Chaired by Lewis, the former was made up of the top physicists and engineers at Chalk River including George Laurence, Ian MacKay, Arthur Ward, Lloyd Elliott, Hugh Carmichael, Don Hurst, and Hank Clayton. At a July meeting of the group, Lewis declared that he wanted to present "a reasoned case for a controlled expansion of the applied research branches of the project" to the board of directors when they met at Chalk River in the fall. In order to prepare, the group reviewed a number of possibilities.

First on the list was a 500-MW (thermal) heavy-water natural-uranium reactor. The possibility of enrichment and the use of liquid metal as coolant was suggested for this reactor. Next in line was a fast fission breeder reactor that would form plutonium from natural uranium. Other projects included a linear accelerator to study particle physics and a high-temperature reactor fuelled by uranium-233.

The organization of the various panels and committees working on future projects was also outlined at the meeting. The basic division was between the Nuclear Physics and Reactor Engineering group and the Metallurgy and Chemical Processing group. Out of these, a group working on thermal neutron power reactors was formed along with a fast-reactor panel.[8]

The program outlined at this meeting was ambitious, to say the least. Research was planned into both fast and thermal reactors, high-energy systems, and particle physics. Despite earlier enthusiasm about the heavy-water natural-uranium reactor, Lewis and his colleagues appeared determined to keep their reactor options open.

At a meeting of the Metallurgy and Chemical Processing group later that July, Lewis reviewed the agenda. Discussion focused on the sale of plutonium, which, it was hoped, would allow the NRU reactor to be a major source of revenue. Although Lewis believed that the price for "bomb"-grade plutonium would drop, there would still be a market for "reactor"-grade material. For this reason, he still hoped that a reactor could be planned to produce both power and plutonium. One possibility was the "recycling power reactor," which would have a core of plutonium or uranium-233 surrounded by thorium or natural uranium.[9]

Lewis participated in all the panel and group discussions, usually as the chairman. His style was to summarize the work done to date and outline future projects. Participation by other members was frequent but Lewis usually had the final say. It was his responsibility, after all, to present the case to the board of directors to ensure that work would continue on the various projects. Most important was that research should continue into power reactors. It was essential that the board be convinced of their economic viability.

Ideas were still ranging far and wide on the actual form of the power reactor. Although at a previous meeting Lewis had suggested that slurries, a homogeneous mixture of uranium and the heavy-water moderator, should no longer be considered, they were discussed again at an August meeting. Advantages such as no sheathing or rod handling, and the possibility of steam generation right inside the reactor, which would reduce costs for heat exchangers, made their consideration worthwhile. As chairman, Lewis was responsible for the unfocused approach of the group. This method reflected, in many ways, his perspective as a scientist who wanted to ensure that every pathway had been closely examined.[10]

Not everyone agreed with this approach. Ian MacKay, a young and creative engineer at the project, argued that in order to build an economical power reactor, it would be "necessary to define the requirements at an early stage in the design more rigidly than for an experimental pile." In other words, MacKay continued, it would have to be "designed for a specific purpose and money [could not] be wasted in making it a 'universal reactor'." MacKay believed that power should be the primary purpose of the reactor and plutonium only a by-product, since the price of power was more stable than that of plutonium. The reactor should not have experimental facilities attached to it and should be located as near as possible to the place where the power would be consumed.[11] It seems likely that MacKay, after witnessing the problems engineers working on NRU were experiencing with the frequent design changes, wanted to avoid repeating these mistakes in building a power reactor.

By September Lewis was ready to go before the board with his list of proposals for future projects. At the top of the list remained the natural-uranium heavy-water power- and plutonium-producing reactor. It would be a 400-MW reactor producing 120 kilograms of plutonium per year and 60,000 kw of electricity. Gone, however, was the fast-neutron breeder reactor. It would appear that MacKay's suggestions were not without impact.

Other requests for funding involved applied research into fluidised fuel (or slurries), research into the electrical generation of fissionable material, development of plutonium-extraction processes, and engineering research facilities to exploit NRX and NRU to determine better reactor design. To pursue these projects, more scientific and technical staff would be needed at the plant, which in turn would require the further expansion of Deep River. If nothing else, Lewis's presentation should have impressed the board with the variety and scope of the work being done at Chalk River.[12]

The board's response was, on the whole, positive. After listening to Lewis's technical descriptions of the projects, the members agreed "that the projects recommended should be further explored, if the financial and other conditions could be met." Mackenzie was not yet willing, however, to commit himself firmly to the necessary funds.[13]

Lewis continued to sell his ideas through lectures and written reports. Early in 1953 he gave a pep talk to the staff at Chalk River. Competition existed in science, he noted, in the form of a race to publish. But if publication was not possible, it was easy to become lazy. He tried to spur the scientists at Chalk River to work harder to fill in the knowledge gaps that existed at the plant. By listing past achievements he encouraged his colleagues to strive for more. Lewis saw future research following two main lines: economic power from fission reactors, and from other nuclear processes as well. The types of fission reactors envisaged were natural uranium, enriched reactors, and fast neutron breeder reactors.[14]

Lewis also had to present his ideas to people who were not trained scientists, such as parliamentary groups. Through lectures he explained to them that despite reports that atomic power was nearly available, in fact many years of hard work still remained. He added that, in most countries, the effort had not been as efficient as possible since much time had been spent on developing weapons. It was essential for people to realize that a lot of effort must be expended before atomic power would be possible. Unlike other types of energy, where the mining is expensive but the burning of the fuel is not, mining of nuclear fuel is cheap but the heat-extraction process is very complicated. Research, he argued, is like the goose whose product is the golden egg. The egg is useless unless man crafts the gold in some way. The value of the "research egg" might not be immediately apparent; but, for example, radioisotopes are now used by a variety of people in many different situations, from the forester who wants to know how sap moves in a tree to the doctor who uses it for cancer therapy. Lewis also examined the "goose," i.e., fundamental research. He argued that, in order to add to knowledge, you have to pursue such research even when it is not clear what use it will be. When discoveries are made in basic science, technological developments will occur. All of this is a step towards the final goals. Perhaps the most significant message Lewis managed to get across was that, despite the practical direction the Chalk River project was taking, it could not lose sight of the vital importance of pure research. Without continued emphasis upon pure research, practical benefits would not be forthcoming.[15]

In the fall of 1953 an Atomic Power Prospects symposium was held at Chalk River. The technical papers concentrated on the potential of atomic power for generating electricity. It was hoped that this gathering would help bring the engineering community up to date and suggest future directions. Subjects ranged from engineering problems in reactor design, to different reactor types and materials, to power planning for Ontario.[16]

This last topic stemmed from the increasing involvement of Ontario Hydro in Chalk River's power reactor program. Ontario Hydro's interest in the possibility of generating electricity from nuclear power dated back to September 1952, when Mackenzie informed C.D. Howe of his invitation to Hydro engineers to form a team to study power reactors. After reviewing the different reactor types, the Hydro team reported that a homogeneous reactor type was the most promising. Lewis did not agree and simply ignored the report. Despite this dismissal of their first proposal, Ontario Hydro engineers continued to work at Chalk River on the power reactor program.[17]

By the end of 1953, the time had come to move beyond the speculative stage and initiate a concrete, detailed study on the possibility of constructing an atomic power plant. Mackenzie had stepped down as president during the summer but he remained involved as a consultant. He joined the new president, W.J. Bennett, at a November meeting with Lorne Gray and R.L. Hearn, the chief engineer of Ontario Hydro. In a report presented to the meeting, Lewis contended that a feasibility study was the next step down the road to an atomic power plant. Considerable judgment would be needed in the early stages of such a study as many technical choices would have to be made. The engineer involved would have to be "not merely technically proficient but capable of assuring the support of those required to contribute to the execution of the plan." For this Lewis suggested a partnership between Ontario Hydro and AECL. Lewis's report was greeted enthusiastically by the others. Hearn stated that Hydro had assigned $200,000 for a two-year feasibility study; AECL, Gray reported, was also prepared to contribute funds. There was some concern that it would appear that Ontario Hydro was receiving preferential treatment but it was agreed that in the publicity it would be announced that Hydro had simply been the first utility to offer its services. The Nuclear Power Group (NPG), as it became known, planned an eight-month study on the construction of an atomic power plant. After completion of this study, it was hoped that a reactor could be constructed in three years.[18]

POWER REACTOR GROUP

Early in 1954, a Power Reactor Group was formed to review work being done on power reactors, including that of the Nuclear Power Group. Chaired by Lewis, the Power Reactor Group would be vital in deciding the characteristics of the proposed reactor. George Laurence, Ian MacKay, Lorne McConnell, Don Hurst, and Hank Clayton were among the staff from Chalk River who participated. Ontario Hydro sent its brightest young engineer, Harold Smith, along with three colleagues. John Foster from Montreal Engineering also took part. There were many forceful personalities in this group, not the least Lewis's. Nor were disputes uncommon. Laurence and Lewis, for example, differed on many aspects of reactor design while MacKay objected to the lack of discipline in their approach. It would prove to be a difficult challenge to balance the various ideas proposed in meetings of this group.

The Power Reactor Group planned to meet each month to discuss the various reactor elements. Lewis had an optimistic schedule in mind; at the first meeting in March 1954 he announced that he hoped they would have "all fundamental characteristics" of the reactor decided by 1 May.[19] It quickly became apparent, however, that this goal could not be met. The group was discussing questions that had never been asked before, and often more than one solution presented itself. With their lack of experience it was difficult for members to decide which was the best answer, and disagreements were common. Scientists from Chalk River were hesitant about fixing certain design specifications early on for fear they would get locked into a less-than-optimal design. Others argued, as MacKay had earlier, that decisions had to be made if they were to move forward.

The overall objective of the group was to design a reactor that could compete with a coal-powered reactor with coal costing $8.50 per ton. To achieve this, every aspect of the reactor's design had to be optimized. At the first meeting of the Power Reactor Group the basic design questions of the reactor were discussed. What type of fuel would be used? Would it be housed in a large pressure vessel or in individual pressure tubes? How would the sheathing material around the fuel react under pressure? Would heavy water be used for coolant as well as moderator? What general form should the reactor take? None of these questions had easy answers and the diversity of opinion soon made it apparent that the design would not be set by Lewis's proposed deadline.

One of the first questions to be tackled by the group was sheathing. The fuel inside the power reactor would have to be sheathed in a material to support the fuel and contain the fission products. The problem was to find a material capable of standing up to the high temperatures, pressures, and radiation fields present inside the reactor. Further, the material could not absorb too many neutrons as neutron economy remained a constant theme throughout. A large part of the problem in making any decision was a lack of information. Zirconium, an element also being tested by the Americans and the British, was a possibility but much remained unknown about its corrosion rate under reactor conditions. Alloys of zirconium were under investigation; aluminum and stainless steel were other possibilities. The latter, however, absorbed too many neutrons and, as Lewis reminded the assembled group, "neutron economy must not be forgotten even in a small reactor."[20]

The question of sheathing continued to top the agenda at subsequent meetings. Despite Lewis's doubts, it was reported that stainless steel appeared to be the best approach in terms of technology, availability, and cost. Lewis continued to push for use of zirconium or a zirconium alloy. The British and Americans had also been investigating the material; Lewis suggested that, instead of performing expensive irradiation studies in NRX, they should rely on the United States for information on zirconium. However, this still did not lead to a final decision. By mid-July 1954, the group was only able to propose that they clad the fuel elements in zirconium, but that stainless steel could not be ruled out.

Even the general shape and configuration of the reactor remained uncertain. Many different ideas were put forward but no suggestions adopted. In April Lewis presented a design for a small, pressurized, 50 MW (thermal) reactor using natural uranium and heavy water. A significant aspect of this design was its assumption that a pressure vessel would be used. The purpose of the pressure vessel was to prevent boiling by keeping the cooling circuit of the reactor under pressure. It would be constructed from a special steel and would likely have to be ordered from a company outside of Canada. Later, the assumption that a pressure vessel should be used for a power reactor would be hotly debated.[21]

By July, it had been determined that the reactor would have a power of between 30 and 50 MW (thermal), fuelled with natural-uranium metal, using heavy water as moderator and either heavy water or light water as coolant. The question of the type of coolant to be chosen provoked extensive discussion. The main advantage of heavy water was its low neutron absorption. However, this had to be weighed against the fact that it was expensive to produce and would, at this point, have to be purchased from the United States or Great Britain. But reports of ongoing studies suggested different methods for producing heavy water. The question remained undecided since Lewis and other scientists at Chalk River wanted to keep open the option of using light water as coolant.[22]

Despite Lewis's hopes for early decisions on design specifications, it became obvious as one meeting followed another that little progress was being made. The problem lay in disagreement over objectives. At the first meeting of the Power Reactor Group, Lorne McConnell had questioned the decision to build a pilot plant first; he suggested that a full-scale plant be designed and constructed. Lewis stressed the need for a pilot plant to demonstrate feasibility and to acquire design experience but noted that the design of a full-size unit could be carried out concurrently with the smaller unit. Doubts persisted, however, since some members believed time and demand favoured construction of a large reactor.[23]

At a meeting at the end of May further objections were raised. In response to Lewis's report on American boiling-light-water reactor experiments, George Laurence complained that they were straying too far from their original objective of building an "NRX-type reactor" to produce power. But Gray and Lewis were still unwilling to limit the study. Percy Dobson of Ontario Hydro agreed, arguing that it was better to explore different reactor possibilities than build a reactor "not much different from NRX." Lewis firmly announced that their overriding objectives were to achieve a reactor that worked and produced power. Harold Smith added that his objective was to see whether the power produced was economic and gain design experience at the same time. Lewis responded that the first power reactor would show the way to cheap power without necessarily being economic itself.[24] Continuing controversy over such fundamentals undoubtedly slowed the committee's work.

When the group met in July, however, Lewis announced that the basic design specifications of the reactor should be laid down at this point. For Lewis, the reasons for building a small power reactor were clear. First and foremost, there was a greater chance of such a project being funded. Next, a small power reactor would provide design and engineering experience for a larger one and allow Chalk River scientists and engineers to obtain information about operation at high temperatures and heat-removal problems.

Other factors influenced Lewis's thinking. He noted that in the United States many different reactor designs were being pursued at the same time with resulting confusion. He also feared that by starting with a large reactor the Canadians might be forced to make hasty, and perhaps incorrect, decisions. On a chart prepared to outline the steps they were taking towards construction of small and large power stations, Lewis noted that they were not stopping work on large power reactors but rather delaying it until more information was acquired. The chart outlined a design study costing $400,000 for the small reactor, to be followed by the detailed design and construction of the reactor, which would be completed early in 1958 at a cost of $15 million. At the same time, design studies would be launched for two large reactors, to be designed and built by 1961 for $60 million.[25]

Reaction to this proposal was mixed. Harold Smith, among others, continued to object to the inclusion of a small reactor, arguing that the sooner the feasibility of a large reactor had been proved, the easier it would be to attract funding for the more expensive program. Others felt that although the small reactor was necessary, too much effort was being expended on it at the expense of the larger reactor program. Part of the problem lay in manpower; many workers at this time were still occupied with NRU and would not be free for a number of years. Lewis agreed that this was a problem and tentatively suggested that a British team might be invited to help.

Despite these criticisms, Lewis remained firm in his belief that a small reactor had to be designed and constructed first. Noting the novelty of the field of reactor design, he argued that with many aspects of the design liable to change, they could not risk building a large reactor first. The experience at Chalk River lay in building research reactors, which were very different from power reactors. Thus, despite the arguments raised against the small reactor, Lewis managed to ensure that it remained the first objective in their program.[26]

The rest of the July meeting was spent trying to pin down the basic design characteristics of this reactor. All that had been decided so far was that it would be fuelled with natural uranium and heavy water would be the moderator. It was noted that using heavy water as coolant would have some important advantages but both Lewis and Laurence wanted to keep this question open, perhaps with an eye to work ongoing in America using light water. Lewis did point out, however, that when the time came to build a large reactor, heavy water would probably prove more economic as a coolant.

It was agreed that the fuel would be long metal rods probably sheathed in zirconium (stainless steel had still not been ruled out). These would be housed in a pressure vessel since stainless-steel pressure tubes had, at this point, been rejected for reasons of neutron economy. With this decided, Lewis asked if there was general agreement on the program as set out and requested that any strong arguments against it be voiced. A few members reiterated their doubts, but eventually Lewis was able to record that "in general there was reasonable agreement that the program was sound and that it should be adopted."[27]

Many committee members might have wondered what they had accomplished since May. Work had been done at a laboratory level and certain design decisions made, but many questions remained unanswered. It was clear, however, that Lewis had achieved his first goal. A small reactor would be built along the lines he favoured before the more difficult task of large-scale power reactors was tackled.

At the end of July Lewis, along with Gray, Laurence, MacKay, and Smith, met with AECL president Bill Bennett, C.J. Mackenzie, and two board members, Geoff Gaherty and Dick Hearn of Ontario Hydro, to convince them that this was the best path to follow to achieve economic generation of nuclear power. Clearly in charge, Lewis led them through the proposal. He readily admitted that there had been many disagreements about various aspects of the program, which was, as a result, only acceptable as a whole. Changing parts of it would not guarantee acceptance. This did not mean, however, as Lewis was quick to add, that there would be no changes as the program progressed. But a definite objective was imperative.[28]

The objective, Lewis explained, was the construction of two large-scale nuclear power reactors. Because much of the cost in design and construction resulted from the novelty of the work, it had been decided to spread these costs over two reactors. They would be fuelled with natural uranium, heavy water would be used as moderator, and each would produce one hundred megawatts (electric)(MWe). A small reactor would not prove nuclear power economic but Lewis postulated that a large one would. In fact, there was no way of knowing until a large reactor had been built and operated.

Although the construction of these reactors was the final objective, the development of the small pilot reactor would come first. This reactor would provide a venue for tests that NRX and NRU, as research rather than power reactors, could not provide. Its construction would result in savings when building the larger reactors, as mistakes made on a small scale would be less costly. Both during and after its construction it could be used to train engineers and operators. Finally, Lewis hoped that by the time the preliminary design was finished, the scientists and engineers working on NRU would be ready to join the small-reactor project and then, once it was completed, move on to design and construction of the large reactors.

Lewis closed the meeting on a note of urgency. The program must be started at once if Canada wanted to retain its competitive edge. If the proposed timetable was not maintained, it was likely that firms in the United States would soon be able to offer better reactors to other countries. Although this was undoubtedly true, Lewis's attitude also reflected his enormous enthusiasm for the project. It combined many of the qualities he loved about science. The reactor program was challenging from both a scientific and engineering perspective. But the goal of large quantities of inexpensive power was what made it most worthwhile. In this project Lewis saw the culmination of all he had been working for at Chalk River. Combining the wonders of physics with expertise in engineering, they would produce a machine that could provide power to Canada and the rest of the world.

The program Lewis laid out was favourably received by the board members, who proceeded to examine it closely. In the next few months they met with the British and Americans to gauge their thoughts on the proposal. Lewis accompanied Bennett, Gaherty, and Hearn to England. Like the Canadians, the British hoped eventually to develop a breeder reactor. In the meantime, they were working on a graphite-moderated gas-cooled reactor. In February 1955, they announced an $840-million (Canadian) twelve-station nuclear power program that made Lewis's proposal seem moderate in comparison.[29] Furthermore, early development work on the Canadian program could be funded by the federal government, but as it grew, utilities would have to contribute. In fact, as Bennett noted, much of the early design work was already covered in the five-year company forecast.[30]

Meanwhile, the work of the Power Reactor Group continued. At its next meeting on 17 August, Lewis explained that agreement concerning Ontario Hydro's involvement in the small-reactor program was not needed until 1956, while 1959 was the earliest date for the large reactor. Much more work was needed in preparing reliable cost estimates since a utility would not want to participate without a more accurate idea of future expenses. Further, as Smith pointed out, more design details had to be determined. The question of using pressure tubes as opposed to a pressure vessel was raised. Tubes were still under consideration since the problems of building a pressure vessel for the large reactor were enormous. However, the pressure vessel resulted in higher fuel ratings, which Lewis favoured. Why discuss pressure tubes for the small reactor at all, then? The question was left unanswered for the time being but would reappear later.[31]

In November, Bennett and Gray joined the Power Reactor Group for a discussion of its work over the previous months. Certain details had been decided: moderator, rod type, zircalloy sheathing, and the general price range were known. But much remained to be worked out and now many more people would be involved in solving these problems. To help coordinate the different aspects of the program, a Power Program Policy Committee was set up in December with Lewis and Gray as co-chairmen. Its role, as Lewis explained, was not to coordinate the other committees but rather to review the coordinative actions of other committees. It would also encompass work still being done on the NRU reactor. Various panels would be set up to examine different aspects of the power program. Finally, it was reported that private firms had been asked to participate in the project and had until February 1955 to present their bids. The board would then decide which bid was most promising and work would continue from there.[32]

One significant change announced at the meeting was Bennett's decision to amalgamate the chemistry, chemical engineering, and process engineering branches at Chalk River into one division to be named Chemistry Research and Development. Questions were raised about moving processing from operations to the new division, but, as the minutes recorded, "it was agreed that any proposed changes to the existing chemical plants would have to be viewed in the light of the needs of the power program."[33] This change, likely suggested by Gray, resulted in power being put before processing as an objective.

Detailed work continued on the small reactor and Lewis remained intimately involved. His overriding priority, neutron economy, influenced every decision. When debate arose in a meeting of the Power Reactor Group over the thickness of the fuel sheathing, Lewis insisted, despite protests from the metallurgists, that they work towards thinner sheathing as this would cut down neutron losses. His tenacity in getting his way on this and other similar subjects has been remarked upon by many who participated in these meetings.[34] They also noted, however, that Lewis was usually correct; in this instance it was possible to make the sheathing thinner.

On 23 March 1955, the federal cabinet formally granted approval for the proposed project. In the meantime, as the Power Reactor Group continued to meet, the features of the planned reactor slowly emerged. At the end of January 1955, a preliminary design had been prepared and presented by John Foster and Harold Smith. It outlined details for the plant and suggested a completion date of 1958. Its planned cost was $15 million and the plant output would be 10 MW (electric). This figure for power rating, as output was called, would be the subject of considerable debate in the next few months and Lewis, naturally, was in the middle of it all.

The question first arose at a meeting at the end of January 1955. Arthur Ward, a senior physicist at Chalk River, suggested that a higher-rated reactor might be more cost effective. But many doubts were raised. It was suggested that design problems might be completely different for a larger reactor. Lewis proposed that they set the planned power rating at 20 MWe with a possibility of decreasing it to 10.[35]

Over the next few weeks, Lewis became convinced that they should build the reactor with the higher power rating and in March he explained his views to Bennett. Noting that a longer period of fuel irradiation in the reactor leads to lower fuel costs, Lewis pointed out that to achieve the longer irradiation, they would have to adopt a fuel configuration that would permit the higher power. If a higher power was possible, he argued, why not work towards it since the company would learn more? Bennett was convinced and wrote to his minister, Howe, explaining why they had made the change.[36]

Although this appeared to settle the question, debate continued in subsequent meetings. In May the first meeting of the Nuclear Power Demonstration (NPD) project committee was held with Canadian General Electric (CGE), the company that had been contracted to build the plant. Their $2-million contribution, which had helped to secure the contract, had been based on the assumption that the rating would be 10 MWe. The problem, as Gray pointed out, was that the larger plant would cost an additional $3 million. Was it worth it? No decision was reached at this meeting but Gray felt the matter had to be resolved by July.[37]

Lewis set forth his views in a paper in early July. Increased capital and operational costs of the larger reactor, he argued, would be offset by the increased revenues from power and plutonium and by the information gained about fuel. A detailed economic argument followed but did little to elucidate the problem. As Gray later pointed out, the question should not be decided on the basis of economic arguments but rather on whether the necessary information could only be gathered from a 20-MWe reactor. This cut through Lewis's complicated and often confusing arguments and was the basis for the final discussion at the next NPD project committee meeting. Lewis apparently convinced the committee that the higher power rating had greater potential. By the end of July Bennett had written to Howe asking for his formal approval for the increase.[38]

As the project progressed Lewis continued in his attempts to convince everybody of Canada's need for nuclear power. Despite his lack of formal training in the subject, Lewis tried to tackle the economic aspects of nuclear power. He had approached this problem a few years earlier in a director's report which he revised in 1955. He predicted a large increase in the use of electric power in the next fifty years in Canada. Further, he noted that the distribution of hydro power was uneven so that highly industrialized sections of the country would have to import coal for energy. By Lewis's calculations, Canada's electrical needs could only be met by sources that would be available in twenty to thirty years and would provide tens of millions of kilowatts of power. Not surprisingly, Lewis argued that only nuclear power could meet these requirements.[39]

Lewis was on shaky ground when trying to prove the case for nuclear power through economic arguments. It was not clear where he obtained his data, and much of his argument rested on pure speculation about future energy requirements. Lewis was much admired for his ability to school himself in new disciplines but it appeared there were limits to his talents.

The question whether nuclear power could compete economically with conventional power sources remained unanswered. This was bound to be an important factor in the political decision to back the new technology. Two reports prepared by the economics branch of the Department of Trade and Commerce examined this question. They concluded that there might be as many as ten large-scale (i.e., 100-MWe) stations in Canada by the early 1970s, but the prediction was based on the assumption that nuclear power could be produced for a price of six mills per kilowatt-hour (one mill equals one-tenth of one cent). If the price rose above seven mills per kwh then there would be "no foreseeable application in this country."[40] If costs could be kept down, the report predicted, nuclear power would still only be useful in Southern Ontario, the eastern Prairies, and the Maritimes since problems in scale and availability of alternative resources would rule out other areas of the country.

A second report on the economics of atomic power reactors was less optimistic about their future in Canada. In markets such as the United Kingdom and Japan, where power was expensive, the outlook for atomic power was more promising than in Canada and the United States. Unless the availability of other kinds of fuel decreased, causing the price of other types of power to rise, the prospects for atomic power plants were not good in North America. Noting that energy sources should not be compared separately but rather as integrated systems, the report still concluded that large atomic plants would be more expensive than hydro when used to supply the base load.[41]

Despite these cautious warnings, both reports still held out hope that the price of atomic power would decrease and, as costs of other forms of power began to rise, become competitive in the future. It was this hopeful projection that Lewis emphasized in reports and lectures on the economics of atomic power plants. In his view, it was only a matter of time before nuclear power could compete economically with other forms of power. Of greater importance was the determination of future demand. Lewis predicted that demand for electricity would rise dramatically in the upcoming decades; to prepare for this, nuclear power must be developed immediately. He repeated this message throughout 1955. It tied in directly with work on the demonstration reactor since only by building a small-scale reactor could scientists and engineers get the numbers necessary for the construction of large-scale reactors. And it was only by building large-scale reactors that atomic power could hope to be economically viable.[42]

The debate over the cost of nuclear-generated electricity was not resolved, but surprisingly, the question was not central to the further development of the demonstration power reactor. Planning began before these questions were asked and it would continue despite the doubts expressed in reports from the Department of Trade and Commerce. From the beginning, the novel and path-breaking nature of the project had proved its strongest selling point. Lewis's enthusiastic backing of it, as well as the dedication of engineers from Ontario Hydro, had helped advance it through rough spots. In many respects, Canadians, like the British and Americans, had been caught up in the excitement of the nuclear age and did not want to get left behind. This, perhaps, was the central reason why the project was pushed forward so quickly when many questions remained unanswered.

NPD

Throughout 1955 and 1956 work continued on the small reactor. In April 1955 it had been decided that the reactor itself would be called the Nuclear Power Demonstration reactor, or NPD. Many of its central features, however, remained open for debate. The question of whether uranium oxide or uranium metal should be used as fuel, for example, was raised at a meeting of the NPD project committee in October. The main drawback of metal fuel was distortion after a limited time in the reactor and at a relatively low temperature. The problems caused by this distortion, such as rupturing fuel cans, had been alleviated to a certain extent by making the fuel rods in different ways. The advantage of uranium metal was that it conducted heat better than uranium oxide, but the latter behaved well at very high temperatures and had much better resistance to corrosion in the event of fuel-sheath failure. At the October meeting, Lewis commented that uranium oxide was looking promising and outlined the work done in that area to date. If the decision was made to switch from uranium metal to uranium oxide, Lewis observed, the size of the pressure vessel would have to increase. The question of fuel type would have to be answered before the vessel could be ordered. This already huge container would be the subject of controversy in the weeks to come.[43]

Lewis's reservations about oxide fuel stemmed from the higher neutron losses associated with it. Yet tests performed by the United States navy in the NRX reactor indicated that oxide could tolerate a very high level of fuel burn-up. This was one of the characteristics Lewis was searching for, and in October he told the CGE team that the fuel would be uranium oxide.[44]

Greater changes would take place in the new year. At a meeting in February 1956, Harold Smith asked the Power Program Policy Committee to consider two new concepts. The first was that pressure tubes be used rather than a pressure vessel, which would give greater flexibility of design. In particular, the size of the reactor could be easily increased. It was also suggested that an enriched thorium oxide fuel be considered. Rumours from the United States suggested this fuel could withstand long irradiation, which would compensate for the increased neutron absorption resulting from adopting pressure tubes.[45]

Lewis supported Smith's suggestions for a number of reasons. These changes would permit construction of any size of reactor. Use of horizontal pressure tubes would minimize congestion and weight-distribution problems and make it possible to use inexpensive aluminum. Gray cautiously asked about the necessary redirection of effort, but Lewis, unable to answer Gray's concerns at that time, suggested that the first step would be to examine closely the problems of pressure tubes. Laurence suggested that they should not shift suddenly from uranium to thorium cycles and his point was noted. Little resulted from the suggested change in fuel cycles, but the first step towards pressure tubes had been taken.[46]

Despite this movement towards pressure tubes, the order for a pressure vessel remained in the works. In June officials from CGE announced that Babcock & Wilcox in Scotland had promised delivery by June 1958. The situation was complicated further when, in December, recommendations were made that a containment shell be added to provide further protection (the design already called for the reactor to be inside a pit with a concrete roof) at an additional cost of $250,000. With the price of the reactor already well beyond what had been initially estimated, this suggestion was greeted with little enthusiasm. But John Foster, who believed that a containment shell was necessary, argued that without one they could not guarantee against leakage for the life of the plant. Gray felt that AECL would be in favour of this suggestion and, as chairman, gave his approval to the expenditure for secondary containment.[47]

The decision to build an extra containment shell did not please everyone. In a letter to the NPD technical committee (which had replaced the project committee), Ernie Siddall, a project engineer, criticized the idea on the grounds that it was expensive and ineffective.[48] Others argued that a containment shell would make inspection very difficult.[49] But Foster responded that the shell would mitigate the consequences of an accident and could not be added as an afterthought.[50]

By early 1957 the NPD project was in a state of confusion. In the previous year a number of questions had been raised. Should the pressure vessel be replaced with pressure tubes? Was an extra containment shell needed? And what would be the final cost of the project? In February 1957, it became clear that the CGE estimates were in a mess and a moratorium was called on the project; further meetings were postponed until CGE had carried out a full review of their estimates.[51]

Although the rising cost of NPD was the major reason behind AECL's decision to re-evaluate the project, another serious concern was that NPD would not adequately fulfil its function as a "demonstration" reactor. In a letter of 25 March 1957, Harold Smith responded to Bennett's request for an update on progress during the moratorium. He noted that the underlying reason for building the demonstration reactor was to prove the feasibility of, and gain experience for, the construction of a large-scale nuclear power plant. Yet after presenting a detailed summary of the work done to date on NPD, Smith concluded that they were about to spend a great deal of money (much more than originally planned) to build a demonstration reactor that would differ significantly from future large-scale reactors. Smith recommended that work be halted so that CGE could redesign NPD along the lines of a large-scale plant. The most radical change he suggested was the switch to pressure tubes as these would be used in the large plant.[52]

Smith's recommendations stemmed from his ongoing work on the large-reactor program. When Lewis had presented his plans for reactor construction back in July 1954, his stated objective had been the construction of two large reactors. Although a full-scale program for the large reactors had not then been put in place, a group of engineers under Harold Smith had started to consider some of the problems involved. As he later recalled, Smith realized that for a large-scale plant a pressure vessel would be impractical; pressure tubes made more sense. Furthermore, the designers would have to move away from the long fuel rods used in the research reactors and replace them with small fuel bundles. The problem remaining was how to convince Lewis that these changes were necessary.[53]

Lewis's dislike of pressure tubes came from his belief that they would absorb too many neutrons. It was for this reason that he had rejected an earlier call for the tubes. But Smith finally convinced Lewis that horizontal pressure tubes would facilitate on-power refuelling, with short fuel bundles being moved in opposite directions in adjacent holes – "bidirectional" fuelling. This would maximize fuel burn-up and help to offset the increased neutron loss due to absorption in the pressure tubes. By the end of March 1957, Lewis was ready to back the pressure-tube idea. He prepared a memo listing the reasons for confidence in the proposed Canadian development program. Heavy water, Lewis noted, had tested well as a moderator in the desired temperature range and could be purchased for twenty-eight dollars per pound. Experience had shown zircalloy to be a corrosion-resistant metal in high-temperature water. It now appeared possible to obtain 8,000 megawatt-days per tonne of uranium without reprocessing and control rods were unnecessary. Finally, there was the "prospect of making satisfactory pressure tubes from zircalloy."[54]

On 27 March 1957, Lewis attended a meeting of the executive committee of the board the review the future of NPD. A new estimate placed the cost of the reactor, including initial fuel and a fuel inventory, at over $24 million. The committee was asked to consider three proposals for NPD's future: the project could continue using the present design; it should be cancelled; or the design should be changed "to incorporate the new design concepts which have been developed in the preliminary design study for the new large reactor."[55] The executive committee, the minutes recorded, decided to pursue the third – Lewis's preferred – option. The committee recognized that the design for NPD and the proposed large reactor had been diverging over the last two years. Since it now appeared that the large reactor would use horizontal pressure tubes rather than a pressure vessel and a have new fuel-loading mechanism, it seemed sensible to incorporate these innovations in the design of NPD. Although the committee understood that this decision would delay the completion of the reactor, it was considered worthwhile.

The March 1957 meeting marked a crucial turning point in the development of NPD. In NPD-2, as the new model was called, the pressure vessel was replaced with horizontal pressure tubes. A bidirectional fuel-loading mechanism, which would become one of the hallmarks of the future CANDU reactors, would be incorporated in both NPD-2 and the design of the large reactors. Although Lewis did not initiate these ideas, once convinced, he backed them wholeheartedly. Foster recalls Lewis telling him about Smith's concept for bidirectional fuelling. This bright idea meant the fuel would be inserted at both sides of the reactor and slowly pushed through, allowing the flux to remain even throughout the reactor. As he gleefully outlined the design to Foster, it was clear that Lewis was thrilled by it.[56]

The divergence that had occurred between the large reactor design and NPD led to a decision to set up a new division to handle their future development together. The Nuclear Power Plant Division (NPPD) would be located in Toronto. This had the advantage of being closer to Ontario Hydro while at the same time maintaining a distance from the scientific community at Chalk River. Chalk River remained involved in the development of the reactors but Ontario Hydro assumed an increasingly dominant role. The large-reactor story and Lewis's role in their development will be outlined in a later chapter.

Work continued on the new version of NPD throughout the fall. In September 1957, a new schedule called for the installation of equipment by spring 1959 and for a reactor to be ready for testing by early 1961. The site of the plant remained the same – up river from Chalk River near Rolphton, Ontario. There was some discussion at the end of the year about plant safety and increased costs. Lewis argued that the safety standards applied to NPD were exaggerated. If the problem were not addressed, he argued, it might "frustrate not only the NPD project but also the whole development of economic power in Canada."[57] Lewis was not arguing that safety standards should be abandoned if their costs proved too high, but he did believe that limits had to be defined. With any reactor there was danger of an accident occurring. But in deciding what steps should be taken to avoid accidents and, if they occurred, to contain them, Lewis argued that the economics of the power reactor should be kept in mind.

This approach was criticized since it appeared to suggest that economic considerations should be put before safety. Indeed, Lewis's comments could at times be read that way. Lewis believed, for example, that a secondary containment shell would not provide the protection its supporters claimed it would; it would simply make the reactor more expensive, power less economical, and the likelihood of establishing power reactors in Canada more remote. Another committee member argued that this perhaps was the price they would have to pay to have power reactors readily accepted by the general public.[58]

In the end it was agreed that construction could continue, even though there was a risk that the AECB would not approve the project once completed. This did not prove to be the case. By the time of its completion in 1962, NPD-2 satisfied the safety requirements of the board and began supplying electricity to the Ontario power grid.

The success of NPD paved the way for the development of CANDU, the commercial reactor design that from the mid-1960s would form the backbone of the Canadian nuclear power industry. All the elements defining the distinctive CANDU reactor type – natural-uranium fuel, heavy-water moderator and coolant, horizontal zircalloy pressure tubes, and a bidirectional fuelling method – were first brought together in NPD. It is therefore worth pausing to reflect on the process that led to this prototype and to assess Lewis's role throughout.

In retrospect, the development of NPD appears to have been both complicated and confused. It did not, as one might have expected, flow logically from conception to design, then construction and completion. Instead, the many different ideas put forward for each aspect of the plant's design were discussed endlessly, discarded and then picked up again, significantly altered in some cases and finally adopted. In many respects, it was not surprising that Canadian scientists and engineers went through this difficult process. They were working in a new area of physics and engineering where many of the variables had to be determined. Other countries experienced similar troubles. The British, for example, examined a number of reactor types before deciding on the one they would pursue.

Yet the scientists and engineers at Chalk River experienced many of the same type of problems in developing NPD that they had when designing and constructing NRU. In part these difficulties stemmed from the scientists' reluctance to discard ideas for fear that they might later prove useful. This attitude frustrated the engineers, who complained about the "reactor-a-day" exercise it engendered. Lewis, in his position as chairman of the most important committees, could have insisted that choices be made and left unchanged. But Lewis was fundamentally a scientist, not an engineer, and he tended to agree with his scientific colleagues that the design should remain flexible until the last possible moment in case a new discovery was made that could improve it. This approach doubtless delayed the final decision to develop a heavy-water natural-uranium power reactor.

That decision does not seem surprising in retrospect; at the time, however, the choice was not as clear. Even after arguing the case for a heavy-water natural-uranium power reactor in his 1951 report, Lewis continued to examine a variety of reactor types. While this reflected his fascination with the subject, more important was his desire to arrive at the optimal power reactor design; the model eventually chosen had first to prove itself against the alternatives. To be sure, the fact that heavy water and natural uranium played to Chalk River's strengths – these had, after all, formed the basis of the two research reactors – gave this alternative an added advantage, one that Lewis was happy to seize upon. But as Lewis's investigations make clear, the final selection was by no means a foregone conclusion.

On balance, Lewis's part in the events that led to NPD and laid the basis for CANDU was both central and positive. His push for the development of a power reactor dated back to his first years as director at Chalk River, and it was Lewis who convinced the AECL board of directors that such a reactor should be built. As chairman of the key technical committees Lewis maintained a close watch on every aspect of reactor design. He questioned scientists and engineers closely on their work, sometimes appearing to know more about their results than they did. Thus, Lewis's role was much the same as it had been during the war when he worked on radar. Instead of introducing new ideas, Lewis synthesized the work of others and harnessed their energies in pursuit of a final goal. He was able to grasp the many different features involved in reactor design and pull them together. Lewis was the driving force that enabled all the constituent parts to coalesce.

The way Lewis worked towards his goals was not always appreciated by his colleagues. Some felt that a team approach was preferable to the confrontational style he adopted. At meetings Lewis would put people on the spot as he challenged their results. Although disliked by some, this method had the advantage of ensuring that participants had a firm grasp of their subjects before presenting them. And once convinced of an idea, Lewis adopted it and supported it tenaciously.

Lewis's methods did not please everyone, but they did produce results. During the 1950s, the scientists and engineers of Chalk River, Ontario Hydro, and Canadian General Electric developed a power reactor that impressed the world. At international conferences throughout the decade, Canada was able to present its work proudly in this advanced-technology area. At the same time, collaboration grew among countries in the area of atomic energy, which resulted in the formation of numerous international scientific committees. Lewis's role as Canada's representative on these committees reinforced his belief in the international nature of science and formed the basis for an important part of his career.

CHAPTER SIX

Canada's Nuclear Statesman

The decade of the 1950s was an important one for the development of nuclear power in Canada. Work continued during this period on the design and construction of a second research reactor for Chalk River, NRU. Despite its high costs and lengthy construction time, the reactor was, when completed in 1957, an important addition to Chalk River. Its predecessor, NRX, remained an excellent research tool after its rehabilitation following the 1952 accident. While the research program at Chalk River advanced, work continued on the power reactor program. After presenting his 1951 proposal to build an atomic power plant, Lewis had convinced the board of AECL to move forward with plans for a demonstration power reactor. By the time it came into operation in 1962, plans for a full-scale power plant would be well underway.

Lewis was deeply involved in both the research and power reactor programs. Increasingly, however, his role expanded beyond the borders of the Chalk River plant to embrace the world of international science. As the scientific director of Canada's only nuclear research laboratory, it was not surprising that Lewis became Canada's nuclear statesman. And it was a role that Lewis accepted with pride.

Collaboration had long been an important part of the scientific process. The wartime secrecy surrounding atomic energy research that had disrupted that custom carried on in the postwar years. The United States, having the most secrets to divulge, was particularly sensitive about giving away atomic information, while Britain was anxious to maintain the links established during the war and the early postwar years. Canada's trump card was the NRX reactor, which continued to provide important information to the US and Great Britain in the 1950s.

GREAT BRITAIN

Lewis's lasting ties with Great Britain are easily understood. Since emigrating to Canada in 1946, he had remained in close contact with his former colleague and counterpart in England, John Cockcroft. In the early postwar years, Cockcroft had pushed for close relations between the British and Canadian atomic energy establishments. A working reactor at Chalk River, research teams, and a continuing British presence at Chalk River made contact both desirable and easy. During the 1950s, this contact would be more formal. Throughout that decade, scientists from Harwell and Chalk River participated in joint technical conferences. As in the late 1940s, much of the information exchanged was basic nuclear data, but as both Lewis and Cockcroft became more interested in power reactors, attempts were made to coordinate their efforts in this sphere as well.

The Anglo-Canadian technical conference of September 1952 had covered specific physical problems suggested by scientists in both countries. At that time, various reactor types for the power programs had also been described. Both countries were investigating power reactors and were anxious to learn from each other. At this early date Lewis was still considering different reactor types, but he already favoured the natural-uranium heavy-water model.[1] He could not understand the British pursuit of a model using graphite as the moderator and disagreed with the British when they declared that the two countries were pursuing "complementary" programs. Lewis wrote to Cockcroft after the conference that "from my point of view I could only regret that lack of heavy water was forcing the UK into a second best development of gas-cooled graphite."[2]

Despite their differences, collaboration continued between the two countries. Chalk River remained very important to the British program. At the 1954 technical conference held in England, Lewis agreed to perform further irradiations for Cockcroft. Since the irradiations were of interest to Canada and did not impose too great a work load on the operating staff, Lewis agreed to do them free of charge. Other requests by the British were usually turned down because Lewis felt little was to be learned from them.[3]

The Canadians wanted to interest the British in a heavy water reactor, while the British had aspirations to establish a closer link with Canada in reactor development. The two sides did not, however, have the same reactor in mind. Christopher Hinton, the engineer in charge of building Britain's atomic energy plants, explained to UK director Lord Plowden that Howe was interested in closer relations, but only on a Canadian national basis. "Heavy water is not now so popular with the AECL directors outside of CR itself," Hinton wrote in September 1954, "and I believe the possibility of a wholly British nuclear project in Canada is improving." The one fly in the ointment was Lewis, who, according to Hinton, was "attached to the heavy water type with an almost bigotted [*sic*] devotion."[4]

At the same September 1954 meeting, Hinton accepted a Canadian proposal to build up a British team to undertake a design study on a full-scale heavy water reactor beginning early in 1955. Although he felt the time scale proposed was unrealistic, Hinton approved of the Canadian program for reactor development. He added that the Canadians should also keep in mind the possibility of purchasing a PIPPA-type (i.e., graphite-moderated and gas-cooled) reactor in three years' time. Acknowledging that the power might not be needed immediately in Canada, Hinton suggested that it would nevertheless be wise to train scientists and engineers to prepare for a future expansion.[5]

The Canadians were in no position to take up this proposal. They were, at this time, experiencing manpower shortages, which in fact partly explained their suggestion that the British set up a design team to work on developing a heavy water reactor. In February 1955 plans were going forward for a UK team to arrive in Canada in May, by which time, however, the British plans had changed. Hinton wrote to Lewis that because of "increased commitments on the military side," they would no longer be able to constitute a heavy water reactor design study. Outlining the full program the British were trying to undertake, Hinton added that it was unlikely that they could fit in a new project in the next three or four years; when the time came, they would have to choose between a heavy- or light-water reactor.[6]

Lewis was unhappy at this turn of events, which he believed to be detrimental to the Canadian program. When discussing the large power reactor program in September, Lewis noted that the British seemed ready to give up completely on water-cooled reactors. He suggested trying to change their minds and added, "I do not feel that the help we may get from the UK should be discarded lightly." Lewis eventually had to admit defeat, however. In September he informed Bennett that alterations would have to be made in their program to take account of the British changes. The work done so far had proved a disappointment. In Lewis's opinion, none of the really good engineers who had worked with Hinton at Risley were on the study team. In fact he believed that the staff at Risley had lost interest in water-cooled reactors. Bennett responded that the British had made it clear to him that they would not be pursuing water reactors; there was little AECL could do about this except perhaps try to build up new connections in the United States.[7]

Collaboration between the British and Canadian atomic energy projects was breaking down in other areas as well. Under previous arrangements between the two countries, Canada had allowed British use of the irradiation facilities at Chalk River. But when the Canadians proposed that the British rent space in the NRU reactor, Hinton expressed reservations. In a letter to Cockcroft, Hinton argued that their past experiences using Canadian facilities had been unsatisfactory. The remote location of Chalk River made it difficult to control the experimental work and evaluate the results. In the future it would be more convenient to carry out irradiations in the Windscale reactor in the United Kingdom. Hinton also doubted that NRU would be finished in time to be useful to the British program. The high flux might be of use for certain experiments, but in general that area of co-operation was no longer as important to the British as it had been.[8]

UNITED STATES

In lieu of a close relationship with the British program, Bennett suggested to Lewis that they try to establish closer contact with the American program. Although this appeared to make good sense, it would prove difficult to carry out. Despite the scientific contacts established during the war, Chalk River did not have the same "special relationship" with American laboratories that it had with Harwell. Lewis, appointed scientific director of Chalk River only in 1946, had not formed close contacts with his American counterparts during the war. His work on radar had been far removed from atomic energy research. And although he had also been out of touch with British work on atomic energy, he knew Cockcroft well both from Cavendish days and from their concurrent work on radar. Lewis's close relationship with Cockcroft made it easy and natural to maintain scientific collaboration in the postwar years; there was no corresponding personal bond with the Americans.

Political circumstances also helped to inhibit the formation of closer ties between Chalk River scientists and their American counterparts. The 1946 Atomic Energy Act (the McMahon Act) severely restricted the dissemination of information on atomic energy research in the United States. Until the act was amended in 1954, interchange between Chalk River and American laboratories would be carefully monitored.

Another factor contributing to the limited amount of collaboration between the two countries was the emphasis in the American program on weapons development. The 1949 explosion of an atomic bomb by the Soviet Union and the American decision a year later to develop the hydrogen bomb further delayed American research on nuclear power reactors – an area that interested Canada. It was not until President Dwight D. Eisenhower's 1953 "Atoms for Peace" speech introduced more openness in the United States atomic energy research program that opportunities for closer collaboration appeared.[9]

Despite these limitations, Canadian-American interchange did occur. It was coordinated through the office of a US liaison officer stationed in Canada, the first of whom, US Army colonel Curtis Nelson, arrived in Chalk River in 1947. Nelson's main task was to look after the exchange of information largely in the form of technical reports. During this period most of the exchange centred on how to manufacture and use heavy water. Later, attention shifted to the NRX reactor.

A 1947 report prepared by Walter Zinn, director of the Argonne National Laboratory in Illinois, outlined the current status of research reactors in the United States. Although six research reactors were operating at various laboratories across the country, five of them were zero-power reactors and the sixth had a low neutron flux.[10] None of them had a very high neutron flux or the capability to test reactor materials and components. The NRX reactor at Chalk River had both.

The United States Navy made early use of the Canadian reactor in their work to design a small reactor for powering a submarine. In the early 1950s NRX was used to test fuel prototypes for the naval reactor. In this instance, despite the secrecy involved, the interchange greatly aided Chalk River scientists. Both the zirconium alloy used to clad the fuel elements, and the type of fuel used (uranium oxide), were seen by the Canadians to be worth adopting for their own program.

Despite the security restrictions imposed by the navy and the McMahon Act, informal collaboration between the Americans and the Canadians occurred. Not surprisingly, it proved impossible for a group of scientists to work together on a set of problems without exchanging ideas. John Melvin, a Chalk River physicist who worked with the Americans during this period, recalled that even as the American physicists kept in daily telephone contact with their US laboratory, Canadian and American scientists worked together as "part of a gang."[11]

After passage of the amended Atomic Energy Act in 1955, a formal agreement on collaboration was put in place between the governments of the United States and Canada. The agreement, dated 2 May 1955, covered collaboration on atomic energy for civilian purposes in Canada, and for both peaceful and defence purposes in the United States. The preamble to the agreement acknowledged that, despite the preoccupation with the use of atomic energy for defence needs, "an increasing number of opportunities exist for the development of the peaceful applications of atomic energy." The agreement spelled out the types of information that could be exchanged. Information on power reactors being developed to propel ships, submarines, and aircraft, for example, would not be exchanged unless a civilian use was apparent. Restricted information on power reactors would also be withheld unless the country wanting the information was developing that type of reactor, or was seriously considering development. Exchange would occur on all matters relating to source materials, isotopes, metals, and compounds; health information would also be freely discussed.[12]

In many ways, this agreement simply formalized the type of exchange that had been going on for many years. Ironically, it was signed at a time when the Canadian and American power programs were beginning to diverge. By 1955 Canada was committed to building heavy water reactors. The Americans, however, had let private industry into the atomic power game and many different reactor types were being pursued. Although interchange with the United States certainly did not cease, there was no longer the need, as the American program diversified and grew, to rely on Chalk River for reactor space.

THE CANADA-INDIA REACTOR

Canada's collaboration with Great Britain and the United States on atomic matters was a continuation of the links that had been established during the war. During the 1950s, while those links remained, Canada also forged new ties with other countries. The most important of these was the partnership formed with a fellow Commonwealth member, India. In 1955 the suggestion was made that Canada offer India an atomic power reactor through the Colombo Plan. A surprising proposal, perhaps, but one that would soon bear fruit.

As a senior member of the Commonwealth, Canada had actively participated in the formation and implementation of the Colombo Plan, a Commonwealth aid program established in 1950 to contribute funds to help underdeveloped nations in Southeast Asia. In part, the plan grew out of a belief that Commonwealth countries should act together to aid others; this was reinforced by the argument that economic help from the West would prevent these countries from being wooed into the Communist bloc.[13] Lester Pearson, then Canada's secretary of state for External Affairs, felt Canada could help encourage India's friendship with the West.[14] Of greater importance from a scientific and personal point of view was the friendship between Lewis and Homi Bhabha, director of the Indian Atomic Energy Commission.

Homi Bhabha came from a prominent family in India. He and Lewis met while they were studying at Cambridge. Both members of Gonville and Caius College, they shared an interest in physics, music, and art. Bhabha performed well at Cambridge and, after receiving his doctorate, returned to India determined to build a nuclear research centre there. After directing a number of scientific institutes he became the head of the Indian Atomic Energy Commission; in 1954 he was named secretary of the new Department of Atomic Energy. A year later he announced that India wanted to build a research reactor. It was at this stage that the Canadians stepped in.[15]

In March 1955 Nik Cavell, administrator of the Colombo Plan, had asked AECL president Bill Bennett if the company could provide India with an atomic power reactor. As a power reactor was not yet operating in Canada, Bennett replied that it would be impossible. But it might be possible for Canada to donate an NRX-type research reactor. Bennett spoke to Lewis about this idea and Lewis enthusiastically wrote to Bhabha that they could duplicate NRX in India and would help to train the operators.[16]

The idea of Canada donating a research reactor to India appealed to Lewis on a number of planes. As a devout member of the Anglican church, Lewis firmly believed that he had a contribution to make to mankind. This contribution could be made through his work on atomic energy. It was his desire to develop the peaceful aspects of atomic energy, Lewis would later state, that had led him to accept the position at Chalk River after the war. The enormously destructive power of atomic energy had been demonstrated; Lewis wanted to exploit its peaceful potential.[17]

Work towards exploiting atomic energy for power production advanced steadily in Canada during the 1950s. Lewis hoped this knowledge could be spread overseas. In his view, electrical power was an essential ingredient for a successful economy. If this new method of generating electrical power could be introduced to the underdeveloped countries of the world, it might have an enormous beneficial effect.

Lewis's desire to share the blessings of atomic power with other countries also tied in with his beliefs about the international nature of science. Trained at the Cavendish with physicists from around the world, Lewis had enjoyed the tradition of sharing scientific knowledge among nations. Atomic energy, however, had been shackled with restrictions from the beginning. It was only in the mid-1950s that these began to be lifted. Lewis hoped that Canada would both exploit and advance this new spirit of international collaboration.

Lewis's personal connections with Bhabha were also significant. The two maintained their friendship from college days right up until Bhabha's untimely death in 1966. It must have been very gratifying for Lewis to be able to combine these various threads in the single offer of a research reactor for India.

Bhabha's initial response to the Canadian offer was cautious. He wondered if the flux of NRX was high enough for Indian needs and added that they had been considering a British model that used enriched fuel. Lewis reminded him that the NRX had an advantage over British reactors in that it had a larger irradiation space.[18] Bhabha discussed the question with members of the Indian Atomic Energy Commission. They were impressed by the generosity of the Canadian offer and the fact that NRX used natural uranium as fuel. This meant that India would not be dependent upon a foreign power for its reactor fuel. Bhabha agreed and in September 1955 Prime Minister Jawaharlal Nehru formally accepted the Canadian offer.[19]

The offer was not made without an eye to the future. It seemed clear that India would soon acquire a nuclear reactor from one country or another. By offering a copy of its own reactor, Canada could demonstrate its desire to establish friendly relations with Asian countries – especially those interested in developing atomic energy for peaceful purposes. It was also clear that Bhabha planned to move on from research reactors to power reactors. If India developed a natural-uranium reactor, then Canada could, in the future, supply the fuel elements.[20]

The question of fuel would later become a point of controversy. The Canadians, aware that the NRX reactor was an efficient producer of plutonium, wanted the spent fuel elements returned to them after they had been in the reactor. The Indians firmly held that they would be responsible for these elements. Construction on the reactor continued and the debate remained unresolved when the reactor went critical in 1960. This issue would later return to haunt the Canadians.

In the summer of 1955, however, enthusiasm for the project ran high. Prime Minister Louis St Laurent sent a message to Prime Minister Nehru on the occasion of the signing of the Canada- India reactor (CIR) agreement, which asserted that the joint project should remind the world of the international origins of atomic science and the possibilities for peaceful development of atomic energy.[21] In August, the first International Conference on the Peaceful Uses of Atomic Energy was held in Vienna, with Bhabha as its president. The possibilities of the atom seemed limitless. By sharing the technology of atomic power, bountiful energy could be brought to countries in desperate need.

There were risks associated with sharing atomic technology, however. Almost as soon as the atomic agreement between Canada and India was announced, officials at the Department of External Affairs in Ottawa began to have second thoughts. They were, after all, donating to India a reactor that had originally been designed as a plutonium-production plant. However, no international agreement on the export of nuclear technology existed at the time. And the Indians adamantly refused to accept voluntary controls or safeguards. Officials from External Affairs and AECL likely comforted themselves with the thought that India would have acquired a reactor from some country, so it might as well be Canada.

Lewis was fully aware of the possibility of NRX-type reactors being used for plutonium production. In a paper published in September 1955, he pointed out that any country with a nuclear power program could build up an inventory of fissile material. Although a chemical-separation plant would be costly to build and would make the power produced expensive, a country could easily produce fissile material. Canada, he commented, must be particularly concerned, as the natural-uranium heavy-water reactor was an efficient producer of plutonium. Lewis wrote this article shortly after the decision to give a NRX reactor to India had been made, but he expressed no concern about India's potential to build an atomic bomb.[22]

Meanwhile, construction on the reactor advanced slowly in Trombay, just outside of Bombay. It had been agreed that Canada would pay for the external costs of the reactor and would also train Indians at Chalk River and provide them with living expenses. A lack of trained staff slowed the advancement of reactor construction. Technical problems stemming from the significant topographical differences between Chalk River and Trombay also had to be solved. For example, the NRX reactor at Chalk River was cooled by fresh water from the Ottawa River, which was never warmer than twenty-one degrees Celsius. At Trombay, there was only hot salt water containing silt, which could not be used directly in the reactor. Instead, a closed fresh-water circuit was used to cool the fuel rods.[23] But despite various technical, manpower, and political problems, the Canada-India reactor went critical in July 1960.

For AECL, the transfer of Canadian reactor technology to India was a success. Canada had joined the other nuclear powers in exporting its nuclear technology. The company also hoped that it had laid the basis for future reactor sales overseas. Lewis also had reason to be pleased. Collaboration with his friend and former classmate Homi Bhabha had led to the transfer of an atomic energy research reactor to an underdeveloped country. In the future, this could lead to atomic power plants and thus, in Lewis's view, greater prosperity. The only element spoiling the picture was the question of safeguards.

It seems likely that, although Lewis realized that the Canada-India reactor would give India the means to build a bomb, he did not believe they would do so. In an article he wrote with Bhabha in 1958, it was stated that "India would ensure that the reactor and any products arising from its use would be employed for peaceful purposes only."[24] Unfortunately, this did not rule out the possibility of a "peaceful" atomic bomb.

It is uncertain what Bhabha's feelings were towards India's possible development of an atomic bomb. French physicist, Bertrand Goldschmidt claimed that Bhabha wanted Nehru to declare India's renunciation of atomic weapons.[25] Nehru felt this option could be explored when India was closer to being capable of building bombs. By that time, both Nehru and Bhabha would be dead.

In 1974, India exploded a "peaceful" nuclear device. Lewis had retired a year earlier and left no written statement of his reaction to this event, but those close to him have since reported that he was very upset by the news.[26] This was the dark side of the nuclear cooperation he had sponsored so readily in the 1950s. In 1974, events had come full circle. Lewis, in trying to promote the positive, peaceful aspects of atomic energy, had become part of an effort that ultimately led to the explosion of an atomic bomb by another country.

ATOMS FOR PEACE

International cooperation on atomic energy matters was taking place in a larger sphere in the early 1950s. In December 1953, President Eisenhower made what subsequently became known as the "Atoms for Peace" speech before the General Assembly of the United Nations. Eisenhower proposed that nations hand over part of their stockpile of fissile material to a new international organization, the International Atomic Energy Agency (IAEA). The IAEA would be responsible for discovering new, peaceful applications of the power of the atom. Eisenhower called upon the scientific community to seek novel medical and agricultural uses for fissile material and make new efforts towards supplying power to energy-starved areas of the world.[27]

Eisenhower's speech reflected his belief that the arms race was out of control and something had to be done to halt it. Although his proposals did little to alleviate the arms race, they did put forward a positive standard for the application of knowledge about atomic energy. Eisenhower hoped that work along these lines could help to make the United States and the Soviet Union partners rather than adversaries in this critical sphere. Such endeavours might also make the smaller nations realize that they had a stake in the peaceful use of the atom. The speech was well calculated to stir the imagination; its impact was felt worldwide.

One concrete result of the "Atoms for Peace" speech was the decision to hold the first International Conference on the Peaceful Uses of Atomic Energy in 1955. A resolution adopted by the United Nations in December 1954 stated that a technical conference would be held "to explore means of developing the peaceful uses of atomic energy through international cooperation and, in particular, to study the development of atomic power."[28] UN secretary general Dag Hammarskjöld, who was responsible for organizing the conference, assembled a scientific advisory committee with members from seven countries. This committee, consisting of representatives from the United States, Soviet Union, United Kingdom, France, Canada, India, and Brazil, would later become the official United Nations Scientific Advisory Committee (UNSAC), which remained in place to advise the secretary general on technical and scientific matters.

Lewis represented Canada on this committee. As head of Canada's nuclear laboratory he was the obvious choice for the position, but the goals of the conference were also particularly well suited to his personal beliefs, beliefs that had led to his enthusiastic participation in the Canada-India reactor: the desire to demonstrate the peaceful aspects of atomic energy, spread this knowledge to underdeveloped nations, and have open discussion on topics that had long been classified. An international conference was an ideal forum for such an interchange. It brought together a large number of atomic experts with widely varying knowledge and experience. It was the kind of international scientific endeavour that Lewis revelled in. Surrounded by scientists who shared his fascination with and belief in atomic power, Lewis was in his element.

The conference also performed a valuable public-relations function. The destructive potential of the atomic bomb had been vividly demonstrated to the world through the devastation of Hiroshima and Nagasaki in August 1945. A decade later, scientists from around the world gathered in Vienna to discuss peaceful aspects of atomic energy. Its utilization for power reactors and medical purposes, among other things, helped to alter the public image of the atom. People wanted to believe this new power could be used to benefit mankind. The conference helped to reaffirm that science was still an international activity. For too many years after the war, as noted earlier, security restrictions had impeded the natural communication between countries on scientific details of atomic energy research. It was hoped that the UN conference would help to break down these barriers.

To a remarkable degree the conference did contribute to the general thaw that was underway in international relations. The Soviet Union, which was slowly beginning to open up after the death of Joseph Stalin in 1953, surprised everyone by contributing a great many papers on topics previously considered classified. The United States responded by releasing additional papers at the conference. This led to a disclosure of important nuclear data that had previously not been shared even among the United States, United Kingdom, and Canada. The Soviet Union also distributed the proceedings of a recent conference on atomic energy held in Moscow. Taking their cue from the two superpowers, other countries adopted this more open approach as well.[29]

Canada's contribution to the conference was substantial. In the Palais des Nations, where the conference was held, individual countries set up exhibits to highlight their achievements in atomic energy research. The Canadian display presented the Canadian reactors – ZEEP, NRX, NRU, and the still-unfinished NPD – through models and illustrations. As Lewis proudly reported, this exhibit "occupied a most privileged position where it must have been seen by every participant."[30] It was an impressive display from a small country that had not previously been known for its scientific expertise.

The success of this first UN conference inspired the world. In the decade since the war had ended, the subject of atomic energy had been shrouded in secrecy. Suddenly, at this meeting, countries were revealing different methods of plutonium extraction, preparation of moderating media, and other previously classified information. This unexpected openness caught the attention of the general public. The press highlighted the positive aspects of atomic energy, especially in terms of power development.

Positive reaction to the conference led to discussions about a second one. Lewis remained on the advisory board, which debated when the next conference should be held and what topics should be covered. It was quickly decided that the conference would be convened in 1958, but the agenda proved problematic. Lewis criticized a draft agenda because it covered too wide a range of topics. He agreed with Cockcroft, the British representative, that the development of nuclear power should be the focus of the conference. Lewis also wanted greater attention paid to the disposal of radioactive materials. Other members, however, were determined that the scope of the conference should remain broad. The chairman, Dag Hammarskjöld, decided finally that although the conference would cover many topics, emphasis would be placed upon nuclear power.[31]

The second conference was successful but participants later argued that the subject matter had been too broad. Certain classified areas, such as thermonuclear-reactor research, had been declassified and openly discussed but the sheer volume of work that was disclosed left many bewildered. Canada contributed nearly four times as many papers as it had at the first conference.

In discussing plans for a third conference, debate arose over who should sponsor it – the United Nations or the newly founded International Atomic Energy Agency. The most important task of the new agency would be to work towards peaceful uses for the growing worldwide stockpile of fissile material. Established under the aegis of the UN, the IAEA began functioning in 1957. Many argued that it should organize the next conference on the peaceful uses of atomic energy. As Doug LePan of the Department of External Affairs advised Lewis, Canada had played an active role in the establishment of the International Atomic Energy Agency and wanted it to be an "effective and vigorous organization." This meant watching to ensure that the United Nations did not encroach upon its jurisdiction.[32] In the end, however, future conferences on the peaceful uses of atomic energy remained where they had started – under the aegis of the United Nations.

There were problems in setting up the new agency. The Soviet Union worried that donating fissile material to a central agency might lead to its dissemination. They asked what technical safeguards would be put in place to ensure that supplying reactors to underdeveloped countries would not lead to proliferation of nuclear weapons. These questions were discussed at a meeting with technical experts in August 1955. Isidor Rabi of the United States outlined a proposal that would guard against these two dangers. It soon became clear that a scientific advisory committee would be needed for this agency also. Debate continued on the exact make-up of this committee and whether UNSAC should simply be used again. In September 1958 a list was put forward containing seven names from seven countries. Canada was not included. As a telegram from Vienna noted, this was not consistent with Canada's position as one of the five atomic powers. The oversight was soon corrected and in October Lewis wrote to accept his nomination to the new committee.[33]

Despite his participation on the scientific advisory committee for the IAEA, and his country's enthusiasm for the new agency, Lewis continued to be primarily interested in the United Nations. In a letter to Norman Robertson, under-secretary of state for External Affairs, he argued that staff of the International Atomic Energy Agency, although technically international civil servants, were in fact never free from their country of origin. In contrast, Lewis believed that rivalries between countries had not existed during the two atomic energy conferences sponsored by the United Nations. Furthermore, he dismissed "any attempt to divorce the United Nations from science" as "both futile and undesirable because of the changed and changing role of science in world affairs and culture."[34]

Not all scientists agreed with Lewis's view of the "role of science in world affairs." NRC President E.W.R. Steacie wrote to Robertson to express his concern at the number of scientific problems being raised at the United Nations. In Steacie's opinion there was a risk that the "dramatic aspects of science will be overstressed and the wrong approach to scientific problems advocated." Instead, these questions should be tackled by international scientific organizations.[35] In a letter to Cockcroft, Lewis outlined Steacie's position and argued that the problem lay in considering the United Nations as simply another international organization, whereas Lewis viewed it as "the supreme manifestation of the limited international cooperation that mankind has so far achieved."[36]

Lewis regarded the scientist's role in the international community as an important one. Although much of a scientist's useful work occurred within his laboratory – the ultimate source of new knowledge – this information had to be shared to be of real value. For Lewis, communication among scientists was an essential aspect of science. Using the example of the Soviet Union, where a combination of isolation and ideology had led to a controversy over genetics (the Lysenko affair), Lewis highlighted the importance of the international exchange of ideas. He believed the United Nations to be an excellent forum for the atomic energy conferences simply because they would have a wider impact than specialized conferences. He urged support for the United Nations' scientific endeavours.[37]

Lewis's involvement with both the United Nations and the International Atomic Energy Agency continued throughout the 1960s. At the end of 1966 he was asked to represent Canada on the UN secretary general's study on nuclear weapons. Secretary General U Thant hoped to prepare a study on the economic, political, and military costs of acquiring nuclear weapons in the hope that this would dissuade smaller countries from obtaining them. Lewis agreed to participate but was reluctant for AECL to become involved any further, since this might detract from AECL's well-known lack of involvement with a nuclear weapons program. Since he had already counselled U Thant on other technical questions, Lewis felt that there would no harm in his helping with the study.

The study group brought together experts from various countries, including Sir Solly Zuckerman from the United Kingdom, Bertrand Goldschmidt from France, and Vikram Sarabhai from India. Members of the group had different views of its goal. One member complained that, as scientists, they were ill equipped to deal with questions best answered by defence and security experts. Lewis was not bothered by this problem. He argued that they had been asked as experts and anybody who felt inadequate to the task should resign. For Lewis, the most important task was to write a report and have it widely distributed.[38]

Lewis was proud of Canada's position as one of the few states that had investigated atomic energy and used it for power purposes without ever building nuclear weapons. He considered that possession of nuclear weapons could only have a negative effect on a country – a view that Goldschmidt could not endorse. Lewis also argued that the Canadian economy had lost nothing by rejecting these weapons; a country could reap the benefits of atomic energy by pursuing nuclear power alone. In pressing this view, Lewis appears to have forgotten that the sale of plutonium to the United States for bombs was essential for the construction of the NRU reactor. Clearly, he did not equate this with participation in an atomic weapons program.

The completed UN committee report appears to have pleased few of the participants. It tried to state the disadvantages to nations of acquiring nuclear weapons: among other things, exclusion from a denuclearized zone, losing access to nuclear material that could be used for peaceful purposes, and forfeiting the economic and political advantages of renouncing nuclear weapons. Lewis's dissatisfaction stemmed from the report's use of threats to stop further acquisition instead of pointing out the benefits brought about by the peaceful use of atomic energy.[39]

Despite this unsatisfactory episode, Lewis enjoyed his role as a "scientific diplomat" with the United Nations and the International Atomic Energy Agency. His belief in both the international nature of science and the positive potential of atomic energy made him certain that international collaboration would continue to grow in the future. Lewis's international activities did not go unrewarded. In 1967 he received, along with Bertrand Goldschmidt (France) and Isidor Rabi (US), two other members of UNSAC, the Atoms for Peace Award. In his acceptance speech, Lewis voiced his belief in "the vision and faith that knowledge of atomic and nuclear science will make it possible to free man and his civilization from any want of his essential needs."[40] This statement explains why Lewis believed that his international scientific activities were the most important of his career.[41]

CHAPTER SEVEN

ING

By the early 1960s the Chalk River project had grown considerably from its fledgling wartime beginnings. It was now the research laboratory of Atomic Energy of Canada Limited, which was involved in isotope production, medical products, and, most significantly, nuclear power. Collaboration with Ontario Hydro and Canadian General Electric during the 1950s had resulted in the design and construction of a demonstration power reactor – NPD. With plans in the works for a full-scale power reactor to be built on the shores of Lake Huron, Chalk River was moving closer to its goal of establishing nuclear power in Canada. Much work on the power reactors remained but their development could no longer be the major objective of Chalk River.

Lewis recognized the changes taking place at Chalk River. He had been the strongest proponent of Canada's development of nuclear power reactors. They had provided a challenging goal around which the company could build its excellent research laboratories. Nevertheless, by the early 1960s Lewis believed that Chalk River needed a new mission. In 1963 he set about determining exactly what that mission would be.

Most important for Lewis was that Chalk River remain in the forefront of nuclear science and engineering research. In the late 1940s and early 1950s, the high flux of the NRX reactor had been the envy of both American and British scientists. And when finally completed, NRU had an even greater flux than NRX. Using these two reactors, scientists at Chalk River had performed first-rate experiments; at the same time, the reactors were used to help develop and design the power reactors. But now, with the power reactor design program winding down, it was time for Chalk River to develop a new experimental device.

With this aim in mind Lewis called together a Future Systems Committee in September 1963 to discuss possible projects and directions the Chalk River laboratory might take. The earlier incarnation of this committee, the Future Systems Group, had puzzled over different reactor types in the late 1940s. The new committee set out to decide on a new project that was feasible and that matched the strengths of Chalk River. Small study groups were set up to examine seven different areas: fusion, fast reactors, proton accelerator, long-distance energy transmission, fuel cells, magneto-hydrodynamics, and direct conversion of heat to electricity. The subcommittees were asked to consider basic principles of the problem, prior work on the topic, its status in Canada and other countries, and its potential for research and economic applications.[1]

In theory, the role of the Future Systems Committee was to examine reports on the seven topics and decide which area should be pursued at Chalk River. In fact, certain topics received more attention than others. In particular, it was clear from the first that the search was biased in favour of a high-flux neutron facility based on a proton accelerator – a topic favoured by Lewis. By December 1963 it was evident that the high-flux facility was also favoured by the committee. It would be a "versatile machine," and "full utilization of all its research possibilities would require a very large supporting staff on the scale of Chalk River services for experimental work."[2]

This decision was communicated to a receptive Lewis. The accelerator would generate protons with an energy of one billion electron volts that would be stopped in a molten target of a lead-bismuth alloy. When the protons struck the target, fast neutrons would be produced. The neutrons would then be slowed down in a surrounding volume of heavy water and collected in beam tubes, where very high fluxes of neutrons would result.

There were many aspects of this project that appealed to Lewis. Conceived on a grand scale, it would easily place Chalk River at the forefront of experimental physics. The intense flux it produced would be many times greater than that available at other leading institutions. Irradiation-type experiments could be performed in the tank while other experiments exploited the neutron beams. Apart from the neutrons, the facility would provide physicists with high-energy protons and many secondary particles, such as mesons. Finally, and most importantly, by using this method of producing neutrons instead of fission, a minimal amount of heat was given off. Unlike a fission reactor, where a great deal of effort is put towards getting rid of excess heat, the proposed facility would only need a simple heat-transport system and the high density of neutrons would still be obtained.[3]

Lewis's interest in a high-flux facility had led him to set up a study group even before the Future Systems Committee was formed. The High Neutron Flux Facility Study Committee met for the first time at the end of July 1963. At first it was optimistically hoped that a machine capable of delivering fluxes of roughly one thousand times greater than NRU could be built within five to seven years. It was the committee's mandate to determine the means of achieving this objective. The first problem was the type of accelerator to be used. The committee at first focused on a linear accelerator, but in August the suggestion of building a separated-orbit cyclotron (SOC) surfaced. The SOC, as discussed in an Oak Ridge National Laboratory report, had certain advantages, including stronger focusing of the proton beam and a lower price tag.[4] Discussions followed with experts in laboratories at Oak Ridge and Argonne. The SOC appeared to be the most promising, especially since advances had been made since the time of the first report on the cyclotron. But problems remained, particularly with respect to costs, which were still far above the maximum that had been set. There was also concern whether the SOC could accelerate the beam to a high enough energy without losing a large portion of it. By April 1964, no firm decision had been taken on accelerator type, but since the SOC was used as the reference design in the studies, it was, presumably, the preferred design.

Discussion in the committee was largely technical as early details about the proposed facility were slowly worked out. The most obvious use for the facility was to provide an intense source of neutrons for experimental research. Details were considered but many questions were left unresolved; for example, it remained unclear whether the beam would be pulsed or continuous. Discussion also centred on the type of target to be used. More neutrons were produced from heavy targets and for this reason uranium was suggested as a possibility. At a later meeting, however, it was realized that too much energy was released from the inevitable fission that took place when a uranium target was used; a bismuth target was proposed instead.[5]

Although experimental research was initially advanced as the purpose of the facility, as early as the first meeting committee members discussed using it to burn uranium-238 and thorium to make other fissionable materials. At first the consensus was that basic research should remain the primary consideration and potential commercial uses of the facility could be dealt with in the future. But members subsequently argued that this topic should be added to the list of those to be investigated. The problem in making economic predictions about the value of fuel produced in this manner was a lack of knowledge about both the value of the product and the capital cost of the facility. At a meeting in September 1963, it was in fact concluded that it would not be economical to produce plutonium with the machine.[6]

It was up to Lewis to ensure that the proposed new facility enjoyed the backing of his superiors. Before presenting the program to the AECL board of directors, Lewis had to sell it to the company's senior management committee. Reaction was not all positive. The chairman, J.L. Gray, believed Canada could and should afford the project, but he questioned whether it should be located at Chalk River. He worried about the considerable expansion that would have to take place there and suggested that the facility should perhaps replace NRU. Furthermore, the support of both the universities and the National Research Council would be necessary to secure the large amounts of federal funds needed. But John Foster and Les Haywood upheld Chalk River as the best location for the project. Another site would slow it down and Chalk River needed to keep up the pace of development in the future. At the close of the meeting it was agreed that Lewis should make his presentation to the board of directors at the end of May.[7]

In May 1964 Lewis made a detailed presentation of the project to the board of directors. Chalk River needed a new focus, he argued. In ten years' time, the engineering development of power reactors might be restricted to supporting an established nuclear industry. A new program should therefore be introduced to allow the laboratory to continue to grow. The scientists at Chalk River had considered trying to increase the flux of NRU but this did not seem to be practicable. A different reaction – spallation – was capable of producing many neutrons, but with minimal output of energy. Lewis reviewed the history of this idea, emphasizing American interest in it as a way of producing fissionable isotopes. US scientists had hoped to use the abundant supply of neutrons to produce plutonium and uranium-235 from supplies of natural uranium and thorium. Lewis acknowledged that the Americans' 1953 calculations for the cost of plutonium was quite high, but added that subsequent advances in technology had probably reduced the price by twenty-five percent. This was still excessive, but Lewis expected plutonium prices to rise in the next decade. In the meantime, he argued, revenue would be generated through isotope production.[8]

It is interesting that Lewis's case for the new project as presented to the board was based almost exclusively on its revenue-making aspects. Clearly, it was important to stress to the board that the extremely expensive project being proposed might be able to pay for itself. Lewis's knowledge and understanding of economics was limited, however, and many of the arguments he put forward were based upon future prices that were impossible to predict. He also trumpeted the potentially profound effect of the project on Canadian industry. Last but not least, Lewis stressed that the accelerator would help to maintain the scientific stature of Chalk River.[9]

Since at the time of his presentation Lewis was not asking for large budget increases, the board of directors agreed that studies of the facility could continue. Lewis still had to convince some of his colleagues that the project was worth pursuing. In presenting it to scientists, Lewis stressed the advancement of scientific research as the central reason for building the accelerator. The high neutron flux would make it possible to do advanced research on the structure and dynamics of solids and liquids and the properties of atomic nuclei. The facility would produce mesons and neutrinos for nuclear matter studies and particle physics research. Lewis also emphasized, as he had done with the directors, that attempts would be made to generate techniques for producing fissile material and isotopes. The development of specific high-tech industries was also cited as a reason for building the accelerator. Finally, and perhaps most importantly, he wanted "to maintain both the scientific stature of Canada and the leading position in research enjoyed at Chalk River."[10]

In June 1964, having received approval from the board, Lewis announced a reorganization of the project into three committees, two of which he would chair himself. The policy committee would be concerned with policy and information related to the accelerator at a higher management level. A technical coordinating committee would orchestrate the efforts from study groups and receive reports from various working parties. Finally, a facility-study committee would be a forum for oral reports and discussion among the representatives from various groups.[11] This organizational structure seemed logical and not too large, but it kept the project internal to Chalk River. By August the facility would have a name: the Intense Neutron Generator, or ING.

The most important decision to be made concerned the choice of accelerator. In the summer of 1964 the favoured model continued to be the separated-orbit cyclotron. The main advantage of the SOC system was the price. By February 1965 a cost-optimization program had attempted to calculate the cheapest way of accelerating the protons. The price varied depending upon the configuration of the SOC, but all configurations were considerably cheaper than a linear accelerator (LINAC). No decision was made at that time as the figures were still considered incomplete. Lewis warned against letting preliminary cost estimates overly influence the direction of the project.[12] By the end of the summer more interest was being shown in the linear accelerator. Although it was clear that it would be more expensive to build and would use more power, there was also greater certainty that it would work. In December it was reported to the policy committee that if the latest ideas were incorporated into the LINAC, then it would be competitive with the SOC reference design. In particular, work at Los Alamos in the United States had shown that in the range of 200–1,000 mega-electron volts (MEV) the LINAC would certainly work. With co-operation from Los Alamos they would be able to build a LINAC more quickly and, once finished, it would be easier to tune and maintain. The figures were briefly reviewed and did indicate that the SOC design would be considerably less expensive, but it was agreed the figures might not be accurate. Lewis decided that since significant advances had been made in the LINAC design, over the next two months equal time should be spent on both projects.[13] This was formalized in February with the formation of a Linear Accelerator Working party. A cost summary for the SOC reference design presented in April estimated the price of the system at $85 million. Of this, over half was due to the complex radio-frequency system. The LINAC design cost-summary estimate was $75 million but the estimate was considered inaccurate and unreliable.[14]

Despite the uncertainty surrounding the LINAC, it was becoming clear that the majority of the policy committee now preferred this machine. It was deemed easier to build and in June Lewis decided that the necessary changes in the program should be made to prepare for a possible shift to the LINAC design. In July 1966, it was officially announced at the policy committee meeting that the reference design would be changed from a SOC to a LINAC. But the LINAC design would not be frozen as improvements could still be made.[15]

When discussing the objectives of building ING, the experimental facilities it would provide were always listed first. But close behind was the more practical objective of isotope production. Lewis had used the production of both fissile material and isotopes in his May 1964 presentation to the board of directors. The importance of isotope production was that it was a means of earning revenue which could in turn help pay for the very expensive machine being proposed.

Commercial Products – the division of AECL involved in isotope production – quickly became interested in the potential of ING for isotope production. At the end of 1964 the division expressed the hope that the new facility could produce up to ten megacuries of cobalt-60 every year. It also anticipated production of other isotopes. Lewis viewed these suggestions as a way of getting the machine to begin to pay for itself.[16] The problem lay in estimating what the revenue would be in the future since it was inaccurate simply to extrapolate from existing markets.

A further difficulty surrounding isotope production lay in deciding its importance in relation to other aspects of ING. In August 1965 the question arose whether cobalt production should be maximized within the machine, or whether its production should only be considered after all other requirements were met. Lewis stated that, while experimental apparatus had first priority in the machine, cobalt-60 production should still be considered an essential feature of the design. He went on to suggest that since cobalt is a good material for neutron capture, using it as part of the shield should be considered. Thorium could be used in the same manner if there was a market for the uranium-233 produced.[17]

Once again, however, the main importance of isotope production lay in the future revenue it would produce. This could also be used as an important selling feature when presenting the ING proposal before government and company boards. In his presentation to the Privy Council committee in August 1966, Lewis stressed in detail the revenue-producing aspects of the machine. He predicted that the ion beam striking the target would produce up to 1.2 grams of neutrons per day, or $16,000 per day, if the cobalt produced could be sold. This could bring in nearly $3 million annually, which could pay for the power bill, a major operating cost.[18]

The enormous expense of ING meant that other methods of selling the project had to be used as well. Isotope production would pay a few bills, but it was expected that ING would have a wider-reaching effect on the economy because of the enormous impact of the design, development, and construction on Canadian industry. It was recognized and discussed early on at a policy committee meeting that the many millions of dollars spent on development for the project should, if possible, be spent in Canada. Lewis agreed that Canadian companies should be approached but warned that, as no government funds had yet been committed, no formal agreements could be made.[19]

Lewis had definite views on the Canadian government's industrial strategy, which he had earlier outlined in his August 1964 memorandum to the board of directors. It was generally accepted, he argued, that Canada needed to increase the research and development component of industry. But with a country the size of Canada, choices had to be made about which industries would benefit from government help. There was little doubt, Lewis wrote, that "the atomic energy field is one that is relevant and ripe for expansion." AECL had already contributed to growth in Canadian industry but there were possibilities for further progress. Lewis drew upon the recently published reports of the Glassco commission on government organization to buttress his arguments. The commission had contended that government support of industry raised employment and tax revenues and also had long-term effects on the industrial capabilities of the nation. In the same vein, AECL, through its power reactor program, had contributed to the eventual lowering of power costs, which would have an immense effect upon Canadian industry. For these reasons Lewis argued that the net capital allocated to operations and research should be increased by twelve percent annually, as it had been from 1952 to 1962. These funds would be put towards a number of ongoing projects at AECL including, of course, the development of the Intense Neutron Generator at Chalk River.[20]

A number of industries were approached early in 1965 and informed that AECL planned to build a machine of an advanced nature and needed specialists to help them. The suggestion was made that staff be seconded to AECL for the eighteen-month feasibility-study period.[21] Unfortunately, many of these approaches proved unsuccessful. RCA Victor Ltd of Montreal, for example, felt it could not contribute since it had no radio-frequency tube-development work in Canada in the area specified. Moreover, they felt there would be no market after ING to make a large expenditure worthwhile.[22]

Lewis nevertheless continued to push his proposals for the use of Canadian industry. At a presentation to government representatives in August 1966, his second specification in a list of five was that "the major effort and expenditure must be in *Canadian* industry, both in the development and production for the world market."[23] Lewis argued that through the design, development, and construction of ING, Canadian industry would be entering into many rapidly growing technological fields. For example the knowledge gained about accelerator construction would allow them to sell accelerators to the universities, which, according to Lewis, were a rapidly growing market.

Lewis did admit, however, that many of the necessary manufacturing skills were simply not available in Canada and that American industry was ready to take orders. This was a well-known problem in Canada and Lewis offered no solution. It is not clear from his remarks at this time whether AECL would have been willing to bear the extra costs and time required to wait for Canadian industry to develop the necessary skills.

Of much greater importance than involving industry in the design and construction of ING was the involvement of scientists and engineers in the university community. Although Chalk River housed a large scientific population, it quickly became clear that scientists and engineers from across the country would have to be involved from the planning stages if a project on the scale of ING were to succeed. The extent of that involvement does not seem to have been immediately grasped at Chalk River. Early planning sessions were completely in-house affairs with no representation from the universities. In April 1965, university, government, and industry representatives were invited to a symposium on ING held at Chalk River. It soon became apparent that the universities would not be content to be informed of progress and then asked to lend their support. University scientists insisted that they should be intimately involved in every stage of the planning, and then with the operation of the new machine. Some months later Lewis acted upon these calls for university participation. In August he announced to his policy committee that there should be a regular channel for dealing with ideas from the universities' staff on ING and its uses. To this end, an ING Study Advisory Committee would be set up with Lewis as the chairman.[24] In September Lewis sent out invitations to fifteen faculty members from universities across the country, as well as to representatives from the National Research Council, the Department of Industry, and the Science Secretariat. The advisory committee's first meeting in November 1965 raised many of the university community's problems and concerns that would return time and again during the next three years.

Lewis had scarcely concluded his words of welcome and introduction before the participants began to voice their concerns. Would ING create too much competition for funds? What would the long-term role of the universities be within the project? What kind of collaboration between the universities and AECL was envisaged for the future? How would the project itself be organized? And finally, where would ING be located – at Chalk River or at a more central location? These were all valid points and will be considered in turn.

The high price tag attached to the ING project worried many physicists. High-energy physicists, who were also reliant upon expensive equipment to pursue their work, were particularly concerned. It was argued that if ING were given a large sum of government money, other facilities would have to go without. If ING could draw upon funds not previously allocated for science, then it would be beneficial; but it should not compete for funds with smaller projects. Lewis argued that such competition would not occur, but few of the sceptics were mollified.[25]

A few physicists on the committee agreed with Lewis. In a memo presented to the committee, Larkin Kerwin of Laval University noted that Canada's research and development budget was still far below the point where ING would freeze out possible competing schemes. Not until there were six or seven institutes the size of ING, he held, would there be a need to worry about competing ideas.[26] Lewis agreed, saying that Canada did not, and should not, have a fixed science budget. At a later meeting, when the price tag attached to ING was nearing $200 million, he claimed that, when spread over a number of years, this was not a large expenditure. By 1975–76, Lewis predicted, research and development expenditure would reach two percent of the gross national product – a sum which would provide for all.

But despite Lewis's hope of an evergrowing R&D budget, there were still concerns that if a large portion of the funds went to ING, other projects would suffer. Lewis claimed that construction of ING would build up industries in Canada with the capabilities to build other major facilities in the future. It was possible, he argued, to avoid destructive competition. But reassurances from Lewis and others at AECL failed to ease the concerns of the academic community. The claim that expending large amounts on ING now would benefit the university scientific community in the future was regarded with scepticism. The issue would return to haunt ING supporters again and again.

Scientists and engineers in the university community were also curious about what their long-term role on the ING project would be. The question of collaboration between Chalk River and the universities was a difficult one that had to be carefully defined. The subject was broached in a letter written by AECL president J.L. Gray in September 1965 to the presidents of the major Canadian universities. Gray argued that ING was not meant to be for AECL's use alone and asked for the active participation of the universities. He noted that some US laboratories, such as Brookhaven, were supported by a number of universities, a practice that could be followed in Canada also. In western Canada the TRIUMF proposal, a cyclotron facility supported by three different universities, was a good starting point, but Gray envisaged another such institute centred on Whiteshell, another in the Toronto area, and one in Quebec. ING, he believed, could be the nucleus of such an institute; universities should be encouraged to participate in its planning.[27]

Gray's letter informed universities of the existence of the ING project and AECL's desire for collaboration but did not explicitly state the form this cooperation would take. Questions about the organization of the project were already being asked. In July Lewis had received a letter from University of Toronto physicist Lynn Trainor asking for clarification on the type of organization planned for ING. Would it be an AECL facility to be used by outside groups, or would it be a national laboratory drawing support from across the country? Trainor argued in favour of a national laboratory but said that AECL would have to coordinate it since the universities lacked the homogeneity to do so. Lewis was not clear in his answer. He argued that, if introduced primarily as an AECL lab, the project would change the nature of AECL. But Lewis saw no need to label the laboratory as national or provincial, or as a Crown company. Everyone should simply work towards an institute of advanced learning.[28]

It was precisely this lack of clarity about the role of the universities that had scientists across the country concerned. At the first meeting of the ING Study Advisory Committee, many specific questions were raised. Larkin Kerwin reported that staff at Laval University were enthusiastic but curious about the universities' role. Lewis responded that he wanted the project to be "of as much use to scientists in universities as is practicable and would encourage the universities to give degrees based on work with ING."[29] But Lewis was unsure what was meant by the term "national laboratory," and the meeting ended with members asking for practical details of how the project would operate and how universities would be involved. Lewis tried to establish that AECL was open to approaches from the universities and stressed that funds were available for visits by university members to AECL.[30]

At the next meeting of the advisory committee in early 1966, Kerwin had prepared a memo for the consideration of committee members. Overall, Kerwin was in favour of the project, but he presented differing views on how it might be organized. At present, he noted, ING was still largely a Chalk River project and should be built even if it remained so. A better solution would be to follow Lewis's suggestion of altering the existing organization to accommodate the universities. University scientists would be encouraged to participate in the design, construction, and research of the facility. But the best solution, Kerwin argued, was that put forward by Gray: a new organization, to be located near Ottawa, Montreal, or Toronto, would be set up with university participation in policy and administration. Kerwin did not believe, however, that this plan was under serious consideration.[31]

Other suggestions were put forward as well. Leon Katz of the University of Saskatchewan suggested that ING be modelled, in part, on the Rutherford High Energy Laboratory, where university-appointed representatives played a key role both in planning scientific investigations and distributing research funds.[32] He also suggested that some of the staff at the project should be based at a nearby university so they could both teach and carry out research. A further important feature of the Rutherford laboratory was the mixture in scientific teams of project workers and university staff. This final point was stressed in the discussion of Katz's memo at the subsequent ING Study Advisory Committee meeting. It was essential to avoid a "visitors" versus "laboratory" mentality at the new facility.

These discussions about collaboration with the universities produced little in the way of concrete results. This was partly due to the uncertainty surrounding the project itself. As more and more time passed without firm confirmation of ING, it seemed premature to work out painstakingly an organizational structure for a nonexistent project. But part of the problem was Lewis's apparent unwillingness to devolve power to the universities. Although he welcomed their involvement in the project, he was hesitant about giving university staff any real control. In his mind, Chalk River was best equipped to design and develop ING in the fastest time and for the lowest price.

Even more controversial than cost and organization was the question of where the project should be located. From the outset, Lewis believed that the best spot for the new facility would be on the grounds of AECL, near either Whiteshell or, preferably, Chalk River. In early discussions, when ING was still an "in-house" project, there was no mention of site; presumably nowhere other than Chalk River was considered. This attitude persisted after the ING policy committee was formed in mid-1964. Since all those working on the project were employed by AECL, it is not surprising that other locations were not considered. At a meeting of the ING policy committee in April 1965, it was decided that members would state that no decision on the future site of ING had been reached. But it would be made clear that the governing factors in the decision would be good foundation, proximity to large engineering services accustomed to dealing with radioactivity, and a radiation-hazard-control group.[33] These considerations were clearly meant to highlight the advantages of locating near the AECL laboratory at Chalk River.

At the end of 1965 the Shawinigan Engineering Company agreed that its engineers would carry out a reference design study for ING. This would include selection of a reference site, preparation of layouts and design, budget cost estimates for construction, operation, and maintenance, and cost estimates for engineering services.[34]

In its initial studies Shawinigan worked on the basis of certain ground rules: the site should be within the borders of AECL property at Chalk River; it would require a large, flat area for construction with a firm bedrock foundation; power and river water should be available at minimal costs; and, finally, it should be accessible to heavy machinery.[35] The assumption that ING would be located near Chalk River was reinforced by the common belief that situating it elsewhere would set the program back by at least two years. It was acknowledged that a Chalk River site had its drawbacks (above all, its remote location), but a significant disadvantage of any other spot was the costs that would be incurred in setting it up. And, as Lewis noted in a draft paper on the location of ING, "this extra cost must be laid at the door of avoiding the Chalk River disadvantage and therefore falls to the account of university and scientific research interests."[36]

This final judgment was appropriate, for it was in the newly formed ING Study Advisory Committee, made up of university scientists, that the first strong calls for a site other than Chalk River were heard. At the committee's second meeting in January 1966, Lewis acknowledged that a site near a large international centre would be ideal but suggested that there would be problems with the large amounts of radioactivity generated by ING. Facilities for dealing with this problem that existed at Chalk River would have to be built at any other location. No decision had yet been made, Lewis claimed; Chalk River was being used as a reference site by Shawinigan simply because they needed a real example to work from.[37]

Calls for a site other than Chalk River persisted, however. At study advisory committee meeting in April, it was suggested that a bicultural choice such as Ottawa, Montreal, Kingston, or St Boniface be considered as French Canadians felt excluded from the work at Chalk River. Bern Sargent, a Queen's University physicist who had worked at Chalk River, emphasized that the Chalk River location was only an "accident of history" and there was no reason for perpetuating it. Lewis responded by initiating a study to determine the penalty involved in using a non-CRNL site.[38]

The resulting report, completed in July 1966, came out strongly in favour of Chalk River. Noting its main advantage – that capital and operating costs would be reduced by using existing facilities – the report estimated that an additional outlay of $2 million to $5 million would be needed and $100,000 to $200,000 per annum if a different location were used. Services such as computer and library facilities, a fire department, hospital, and radioactive-machine shops would have to be built. Further, the land and exclusion area around ING would have to be purchased. If placed near a population centre the facility would need a large and, of course, expensive containment building. All these things would increase the price of ING. The clear conclusion of the report was that ING should be constructed at Chalk River.[39]

Despite these arguments, many scientists in the academic community still argued strongly against the Chalk River location. In fact, as the debate over ING gathered steam, more people came out against the site. In August 1967 a long article appeared in a Montreal newspaper calling for the location of ING in Quebec. Its French Canadian author noted that promoters of ING emphasized the enormous number of scientific, economic, and industrial effects the new laboratory would have. Since most federal laboratories usually ended up in Ontario, however, it was to that province that most of the benefits flowed. The federal government should therefore place the new laboratory in Quebec so that it could enjoy some federal largesse.[40] The question of site had clearly developed into a political controversy. By the end of the year a trip report on the issue confirmed that universities in every province hoped to have ING located near them.[41]

This controversy did not help the overall problem of obtaining government approval for the project. Early in 1967, AECL president J.L. Gray warned Lewis of the importance of the site question. Noting that the board of directors thought there was a good chance that ING would be approved, they asked Lewis not to discuss the question of site until the project's scientific and technical merits had been endorsed. Gray even bluntly asked Lewis not to discuss the question of site at all outside the company.[42] Yet the location issue was only the most prominent of the academic community's many concerns. Equally important was the question of cost.

Estimates of the costs involved in the design, development, and construction of ING had been rising steadily since the project's inception. At the final meeting of the High Neutron Flux Facility Study Committee in June 1964, when Lewis had announced an expansion of the project, he set a limit of $25 million for the accelerator – the big-ticket item. A little over a year later, at the first study advisory committee meeting, Lewis expressed his hope that the price tag would be closer to $50 million than $100 million. Twelve months later a more accurate – and larger – estimate had been determined. Assuming a site near Chalk River, the cost of the new facility had now reached $130 million.

The ballooning cost of the project did not inspire confidence in the academic community. Their central fear was that federal money spent on ING would be money *not* spent on their university science departments. Many worried that ING would simply take up too large a fraction of Canada's investment in science projects. Lewis categorically disagreed with this viewpoint. He denied that Canada had a fixed science budget at all. Although ING's $200 million price tag (a November 1966 figure) might seem large and threatening to university research budgets, it should not be seen that way. Lewis argued that if Canada increased her research and development expenditure to three percent of the GNP (which was, in his view, the minimum expenditure for the continued success of a technologically advanced country), then the expenditure on ING would not seem at all large. In fact, a great many new projects would be possible.[43]

The Liberal government of Lester Pearson was at a loss to decide upon this controversial subject. In September 1966 it turned to the Science Council of Canada, created only four months earlier, for advice. The role of the Science Council was "to assess in a comprehensive manner Canada's scientific and technological resources, requirements and potentialities."[44] It was also to advise the government on science policy, possible long-term objectives, and the best means for achieving them. The members sitting on the council were a mixture of representatives from the government sector, universities, business, and industry. The main disadvantage of the Science Council was its lack of budgetary control. As later pointed out to the Senate committee examining science policy in Canada, this shortcoming was "analogous to a cabinet meeting without the Minister of Finance. It can very easily agree on all of the projects coming from the various departments."[45] Nonetheless, it was hoped that the Science Council could advise the government on the complicated questions of science and technology that kept arising. The first on the agenda was ING.

A committee of eight men, drawn from government, universities, and industry, was formed to examine the AECL proposal. Five formal meetings were held and the advice of experts from the United States was sought. The committee assessed the proposal on the grounds of its scientific, technical, and economic merits. In its report, presented to the Science Council six months later in March 1967 and subsequently endorsed by it, the committee announced that it had found the proposal "well-conceived and imaginative." Its benefits would "extend over a broad range of sciences and technologies" and, in turn, would involve scientists and engineers from government, universities, and industries.[46] The committee acknowledged that the costs of the project were high but argued that, barring sharp escalation, they remained within the bounds possible for the Canadian economy. On one of the more controversial questions of the proposal – that of site – the committee differed from the majority of opinion at Chalk River. Anxious to ensure university and industry involvement, its members recommended that a location with greater accessibility than Chalk River be chosen.[47]

In the end, the Science Council's final recommendation to the government was cautious; a further feasibility study should be undertaken and the project reviewed at the end of it. Provided that costs had not risen excessively and that the goals of the project had not altered significantly, the Council recommended that the project should then be allowed to proceed.

Despite this recommendation, the government did nothing. As scientists across the country waited for government to act, controversy mounted. In July 1967, shortly after the Science Council's endorsement of ING was made public, an editorial appeared in the *Globe and Mail* questioning such a large government expenditure on a single project. Noting that the operating costs for ING were estimated at more than $20 million a year, the writer pointed out that the same sum would support about two thousand doctoral-research students. The editorial concluded that the government should consider whether "the well-being of all Canadians would be better served if the money was spread over a much wider area of scientific research."[48]

Discussion also continued in the scientific journals. In February 1968, *Science Forum* devoted a large part of its first issue to the ING question. In a five-page, closely argued article, Lewis, with two of his senior colleagues, defended ING against its detractors. They particularly stressed areas where they had often been attacked. The article argued against the view that ING would be valuable mainly to the nuclear physicists at Chalk River and instead listed a series of subjects, from space technology to oceanography, where the benefits of ING would be felt. Canada would also gain skills in many areas of high technology, a host of scientific advantages, as well as the prestige of owning and operating a unique and powerful experimental tool. In keeping with Chalk River's original and successful mandate to develop nuclear power, the article also pointed out that the machine could be used to develop the next line of power reactors. Finally, in response to the criticism coming from the universities, Lewis and his colleagues argued that, with university and technical college enrolment on the increase, Canada had to provide work opportunities for the new graduates. In their opinion, the strong interaction that already existed between Chalk River and the university community would only be enhanced by the proposed project.[49]

Many scientists and engineers in the university community did not agree. J. Gordon Parr, responding directly to Lewis, wrote that ING was "the wrong thing, in the wrong place, at the wrong time." There was, Parr argued, more important work to be done. Further, too much money was spent on government in-house research already, and more nuclear research was not necessary until the knowledge already gained from it had been utilized. Part of the problem, Parr noted, was that AECL had performed its mandate too well and thereby worked itself out of a job. Parr felt that the academic communities had not been consulted.[50] Other authors agreed. Ken McNeill, a physicist at the University of Toronto, wrote that there had not been enough discussion with the university scientists concerning the aims and priorities of ING. As he put it, there was "not enough clearing of the clouds around Olympus."[51]

This final statement summed up the sentiments of many scientists in the academic community. Although ING was likely a viable scientific project for Canada – one that would enhance its status in the world and possibly bring benefits to Canadian industry and further knowledge on many fronts – many Canadian scientists were irked by the lack of consultation between them and their counterparts at Chalk River. As J.G. Parr wrote, they were told about the project after it was well advanced and, upon expressing concern, accused of having shown indifference and hostility towards ING.[52] Since the project demanded a massive influx of federal funds, this failure to rally scientists on a nationwide basis was crucial. Although many other problems with the proposal were pinpointed – the isolated site of Chalk River, the overemphasis on nuclear science, the huge budget – the lack of real consultation and involvement with universities on a national basis appears to have been Chalk River's largest strategic error. And in the end it was probably this mistake that killed the project.

At the end of 1967, however, it was still alive. As late as December, the applied physics division was provided with $220,000 for ING development work. But without a positive endorsement from the federal government the future of the project remained in doubt. When Pearson announced his decision to step down as leader of the Liberal party early in 1968, it was clear that a decision on ING would not be made until a new leader was in place.

Controversy continued in the press throughout the summer of 1968. A letter to the editor from members of the National Committee of Deans of Engineering and Applied Science made it clear that many members of the scientific community remained very unhappy with the project. They argued that "if $155 million is to be spent, it would be much better invested in the identification of pressing technological objectives, the establishment of research priorities, and the funding of a number of centres of excellence for the solution of problems that will have a predictable impact upon the Canadian economy."[53]

The news that the project had been cancelled was no surprise when it came in September 1968. In his announcement to company employees, Lewis admitted he was "personally most disappointed" with the decision. He added that while Chalk River still retained a research program in nuclear science, work on ING would be discontinued.[54]

Other colleagues held out the hope that certain aspects of the ING program would be retained. In a memo to the president, Les Haywood pointed out that many continued "to regard the electro-nuclear generation of neutrons to be of considerable importance." He advised that they continue work on accelerator technology, which would leave them in a position "to produce relatively quickly a device for the electric production of neutrons which would in turn permit production of fissile material or power without the initial presence of fissile material."[55] Lewis also maintained the hope that the project could be revived. In a senior management committee meeting convened shortly after the project was cancelled, Lewis optimistically suggested that the project itself might be salvaged in two years time. For the time being, however, ING was dead.[56]

ING was not the only "big science" proposal to be cut by the newly elected Liberal government. Since the early 1960s, a proposal had been in the works for a new astronomical observatory to be located in British Columbia. But as with ING, controversy erupted within the astronomy community over both the price of the new telescope and its proposed location. Confronted with a divided group of scientists, the politicians opted for the easiest route: cancellation.[57]

The abandonment of both these projects can in part be viewed in a broader North American context. In the postwar years, basic science in Canada and the United States received a great deal of financial and popular support. There existed a widespread belief that support of science would lead to solutions to many of the world's problems. This is perhaps best exemplified by the popular support given to the peaceful uses of atomic energy during the 1950s. That support started to wane during the 1960s. People began to realize that science could not solve all of the world's problems and, in fact, seemed to be creating many of them. As doubts grew, people questioned the massive amount of government money that was funding esoteric science projects with no clear-cut purpose. Reacting to this growing lack of public support, politicians responded by cutting back on expensive basic-science projects.[58]

But the cancellation of ING did not stem entirely from a disillusionment with big science. The events surrounding the promotion and subsequent cancellation of ING demonstrated the lack of political acumen of scientists at Chalk River, in particular their leader, Lewis. Convinced of the scientific importance of the ING proposal both for Chalk River and for scientists throughout Canada, Lewis and his colleagues did not work hard enough to ensure the support of university scientists and engineers. Those scientists and engineers, angry at the lack of consultation with Chalk River and afraid that the cost of ING would bite deeply into their own science budgets, opposed the project.

The lack of enthusiasm within the university physics community for the proposed project may also have been influenced by the failure a decade earlier, to win approval for their own "big science" proposal. In October 1958 the Canadian Association of Physicists had submitted a brief to the government requesting funding for a high-energy laboratory for Canada. Six months later the government rejected the proposal. Still smarting from this refusal and concerned about the future of academic physics across the country, it is perhaps no surprise that university scientists were doubtful about the more expensive Chalk River facility.

The controversy surrounding both these proposals illustrates the growing resentment felt by academic scientists towards their government colleagues. As one author put it, the "ING affair showed more clearly than ever before the changing balance of power between the academic and governmental sectors of the Canadian scientific community."[59] The Glassco commission had been the first to argue that government performed too large a proportion of research in Canada. This view would be echoed by the special Senate committee set up in the late 1960s to examine Canadian science policy. Yet the question remained whether a decrease in government-performed research would result in increased support for universities.

The debate over the Intense Neutron Generator highlighted the lack of adequate science-policy machinery in the federal government. It was in the midst of the ING debate that the Senate named a special committee to look into science policy in Canada. Chaired by Senator Maurice Lamontagne, the committee, it was hoped, would pinpoint problems and suggest solutions to ameliorate the confused situation. During its hearings, one witness pointed out that ING was an "excellent case history of how science policy should not be made." After many years of government support, and after the government-appointed Science Council had recommended its continuation, the project had been abruptly cancelled for financial reasons. As one senator put it, there was little point in making priorities for decision making if they were to be disregarded. The government could not simply announce that there was not enough money for ING; it also had to admit that it had decided not to expand the country's science horizons.[60]

In later years, Lewis did not accept any blame for the cancellation of the Intense Neutron Generator. In a letter written shortly before his retirement in 1973, he stated that the decision to abandon ING came about because "some of the Deans of Engineering at Canadian universities did not consider the development of a high power accelerator as a fit engineering project and worked to divide the scientific community."[61] But despite Lewis's denial of his own role in "dividing the scientific community," there is little doubt that the failure to obtain funding for ING was the first major setback in his working life. Sadly, after his many successes, it was ING's cancellation that crowned Lewis's career.

CHAPTER EIGHT

The Power Program Revisited: 1959–73

The decision in 1968 to cancel the Intense Neutron Generator project brought to an end Lewis's drive to develop a new research facility at Chalk River. Although its cancellation was a significant setback, ING was only one part of Lewis's work during this period. His involvement in international science had grown considerably since the first Geneva conference and he continued to be deeply involved in power reactor development. The successful completion in 1962 of a nuclear power demonstration reactor, NPD, was followed a short nine years later by the unveiling of two full-scale power reactors at Pickering. This rapid expansion of the power reactor program was due to a productive collaboration between AECL and Ontario Hydro, for it was in Ontario that most of the power reactors were located. The sequence of reactor construction that occurred after 1962 was as follows.[1]

Plans for a full-scale power reactor began three years before NPD came on line in 1962. The type of reactor had already been determined: it would be a two-hundred-megawatt (electric) natural-uranium reactor with the fuel in horizontal pressure tubes. The Nuclear Power Plant Division (NPPD), which comprised scientists and engineers from both Ontario Hydro and AECL, would design the new reactor. When completed, the plant would be owned by AECL.

After examining various possible sites for the reactor, Hydro purchased land at Douglas Point on Lake Huron in June 1959. Construction began in early 1960; at the same time, design work on the reactor was continuing at NPPD in Toronto. Some experience could be gained from the ongoing work on NPD, although since NPD had yet to be completed, certain questions remained unanswered. Furthermore, the reactor planned for Douglas Point would be much larger than the demonstration reactor, and more complex.

Douglas Point experienced many problems during the course of its construction. While the project did fall behind schedule, it managed to remain more or less within its budget. But when the reactor went into operation in November 1966, the cost of the power produced was considerably higher than expected. Although the reactor was not intended to be economical, the gap between the price of the power produced at Douglas Point and that produced at Hydro's conventional generating stations was larger than anticipated.

Douglas Point remained in operation for twenty years. Throughout that time it was plagued with difficulties, but the experience gained from its construction and operation could be applied to work on a new and larger reactor – Pickering. Once again, however, the decision to proceed with the design and construction of the Pickering reactors occurred long before the Douglas Point reactor had proven itself. In part, this resulted from a concern that, unless a new reactor was commissioned, the design team assembled at NPPD would begin to disperse.

The question of the future of NPPD arose at an AECL executive committee meeting in the spring of 1963. It was known that Ontario Hydro was interested in expanding to five-hundred-megawatt reactors and discussions between the two companies resulted in an agreement by June. The following year, NPPD was housed permanently at a site in Mississauga outside Toronto and in August 1964 it was formally announced that Pickering would proceed.

The Pickering arrangements were different from those for Douglas Point. Ontario Hydro would own Pickering, which would feed into Hydro's basic power supply. Hydro would use AECL, through NPPD, as a consulting engineer. The story behind the construction of Pickering has been well told elsewhere.[2] Despite a number of setbacks, Pickering One and Two began operating in 1971 and were followed over the next two years by the completion of Pickering Three and Four. Unlike Douglas Point, these reactors have proven to be sturdy and have operated efficiently and safely for many years.

The final introduction of nuclear power reactors during this decade occurred in Quebec. Although the province possessed abundant hydroelectric power, there was concern that it would not be enough for the future. Furthermore, there was a desire to benefit from the latest technological breakthrough. In the summer of 1964, AECL and Hydro-Quebec agreed to consider the construction of a 250-megawatt (electric) plant. Quebec's interest in a power reactor coincided with a strong belief in Ontario in the CANDU design undermined by a shortage of heavy water. It was quite natural, therefore, that AECL should suggest trying one of the new designs emerging at that time from Chalk River. It was this coincidence of events that led to Quebec's obtaining a heavy-water-moderated boiling-light-water-cooled reactor (CANDU-BLW). A study was undertaken at Chalk River, with the decision to build coming in 1966.

The expansion of power reactor development in the 1960s brought many changes to AECL. The company progressed from building a prototype power reactor, to a full-scale CANDU power reactor, to a different version using boiling light water as coolant instead of heavy water. Much of the work on these reactors was done by the Nuclear Power Plant Division of AECL in close collaboration with Ontario Hydro. Lewis's deep involvement came about because the decision to build NPD, and subsequent reactors, meant that an immense amount of research and development had to be carried out. In particular, Lewis was closely involved at a technical level with work on uranium oxide fuel and zirconium.[3]

Lewis had always combined an intense interest in detail (he was, after all, an experimental physicist) with wide-ranging explorations of theoretically possible reactor systems. In the early days he focused on the breeder reactor, but he also examined different fuel systems, coolants, and moderators. It is interesting to follow these investigations as they indicate the depth of Lewis's involvement with the development of different reactor systems.

In the late 1950s scientists and engineers at Chalk River were examining a different type of coolant – organic coolant. Lewis was interested in these substances because they had several advantages over heavy water. Organic liquids of the terphenyl class could be used at higher temperatures than heavy water: this would increase the power-generation efficiency. Added to this, the price of organic coolants was low which further reduced the price of the power produced.

Ian MacKay had considered organic coolants when working at Chalk River in the early 1950s. After his move to Canadian General Electric in 1955, he was allowed to pursue this interest and a study of these materials was carried out. The results of this work, along with that of scientists from various other countries, were presented at the second United Nations Conference on the Peaceful Uses of Atomic Energy, held in September 1958. Lewis was intimately involved with the United Nations conferences and would have studied these papers closely in advance of the event. A few months later, he proposed that Chalk River should examine organic coolants in greater depth.[4] Lewis brought up the question of organic reactors at a board meeting in late 1958. Noting that CGE had done some work on organically cooled reactors and that experiments using the substance had been performed at Chalk River, he added that this type of reactor seemed ideally suited for plants of 150-MW (electric) size or smaller. Later in the same meeting it was agreed that work on organic coolants should continue at Chalk River; in fact, work on gas cooling would end and be replaced by work on organic and direct steam cooling.[5]

By February 1959, Lewis and Gray had prepared a firm proposal for an organically cooled reactor – a proposal that fit in well with the political climate. In 1957 John Diefenbaker and the Conservatives had come to power in Ottawa. In an election the following year, they consolidated their position by winning a crushing majority. In both election campaigns the Conservatives had successfully promoted the idea of northern development. Diefenbaker had a "dream of opening Canada to its polar reaches." To do so, Lewis and Gray must have considered, power would be necessary. The timing was perfect for Chalk River to propose development of a small reactor that would be ideal for use in small northern communities.[6]

Gray and Lewis made their proposal in a letter written to the board of directors in February 1959. They described the promising results from experiments on organic coolants. Their optimism about these liquids was backed by the United States Atomic Energy Commission. Work on NPD-2 and the CANDU reactors was progressing well, Gray added, but many at Chalk River believed it was time to add a "second string" to their program. Further, Gray believed that "the general public, the uninformed press, some manufacturing organizations and the uranium producers" were pushing "in varying degrees to expand the reactor construction programme and towards a general increase in the tempo of our programme."[7]

Diefenbaker's "Northern Vision" was carefully tied into the proposed new program. Gray noted that the Department of Northern Affairs and National Resources believed that one of the aims of AECL should be development of a small reactor for use in remote northern sites. This would be an important selling point of the proposed new reactor program. It was recommended that CGE work with NPPD to investigate construction of a forty-megawatt (thermal) organically cooled reactor. The reactor would be built at Chalk River. It should be adaptable to northern conditions and capable of using natural-uranium fuel and operating at the high temperatures necessary for economical power production.

The response to this proposal was positive. At a March meeting, the board agreed that CGE should be given the contract for a design study for an organically cooled, heavy-water-moderated experimental reactor. Over the next few months CGE progressed with its study. Lewis reported continuing positive interest in organic coolants by members of Euratom, the European Atomic Energy Community, which comprised Belgium, France, Luxembourg, the Netherlands, Italy, and West Germany. They had decided to centre their technical program on the same type of reactor as that being developed by CGE so there would be opportunity for future collaboration.[8]

By July, however, a new idea had arisen. In a letter to his new minister, Gordon Churchill, Gray suggested that, with a staff of 2,300 people, Chalk River was "near the saturation point for major facilities and we should be considering a new establishment." Gray proposed that a new research laboratory be established in Manitoba to develop the organically cooled reactor.[9] The fact that Churchill's home was in Manitoba no doubt influenced this choice. The proposal was approved quickly by cabinet and plans for the construction of the new Whiteshell Nuclear Research Establishment east of Winnipeg went forward. Despite initial cost overruns and a modification of the project in mid-stream, by 1965 a successful test reactor was constructed at Whiteshell.[10]

The choice of Whiteshell suited the government's priorities but distanced Lewis from the project. He, not surprisingly, had hoped that development of the organic reactor would take place at Chalk River. Yet Lewis's interest in organic coolants remained, particularly the possibility of their application to large power reactors. An organically cooled CANDU was only one variation considered by Lewis and other Chalk River scientists during the 1960s. In February 1962, the Power Reactor Development Programme Evaluation Committee (PRDPEC) was formed to examine different types of heavy-water-moderated reactors and direct development of the most promising variants. The committee would advise when a particular reactor type ceased to be economical and should be abandoned and would also initiate studies of new approaches. Lewis, as chairman, would report to the president and board of directors of AECL on the different programs and their prospects.[11]

At the first meeting, committee members discussed four different types of CANDU reactors. Two of the designs, the improved CANDU and the fog-cooled reactor, were already under examination and progress reports were submitted. The improved CANDU would build upon the knowledge gained from operating the two research reactors, and from the construction of NPD. But there were hopes that the output of the reactor could be improved by examining new fuels and sheathing. The fog-cooled reactor had the advantage of using light water as its coolant, but it was reported that a great deal of development work still remained.

With the help of a young physicist named George Pon, Lewis proposed to examine another reactor type that would use light water as coolant and heavy water as moderator – a boiling-water-cooled reactor. It was proposed that the steam generated from the boiling-water coolant be used as the working fluid in the turbine. This, it was suggested, would lead to higher plant efficiency and a reduction in heat-transfer equipment. Finally, investigations would continue into an organically cooled reactor.[12]

In March 1963, Lewis presented a report that examined the overall possibilities of power-reactor development. Lewis's analysis of the economics of the situation was strongly positive. Not only were major cost reductions effected both by increasing a plant's capacity and by building more units but the cost savings from development "would make it worthwhile to develop each or all of the [reactor] types considered."[13] And from the figures the committee had gathered, none of the reactor types showed a significant cost advantage. There could thus be no doubt, Lewis argued, that nuclear power would be significantly cheaper than coal. His figures showed that its utilization would save $49 million per year. This was based on the assumption that capital costs for nuclear plants would continue to fall. Lewis predicted that Canada's power needs would grow at a rate of 5.5 percent to 7 percent per year, which meant a doubling of capacity after ten to thirteen years. Nuclear power, he contended, would fill much of this capacity.

For continued savings into the future, however, Lewis strongly recommended development beyond the CANDU. One of the several variations considered by PRDPEC should be developed. In comparing the costs of the different types, Lewis noted that the main reason for varying costs was the fuel type. Heavy water posed a constant problem in that it had to be kept isolated and uncontaminated. Overall, the organically cooled reactor appeared to be the most economical. But since many significant technical details remained to be worked out, this advantage could not be taken as definite. Nevertheless Lewis still argued that "the economic attraction of the organic coolant ... remains a cogent argument for continued development."[14] Lewis did not see the need for significant new expenditures on development; the rate of expansion over the past few years had allowed the program to grow steadily. But he did caution against changing the basic CANDU design significantly, for this would add considerably to total development costs. As an example, he noted that changing from 3.25-inch pressure tubes to a larger diameter would involve development work estimated at millions of dollars. This question of altering the pressure tube size would arise later and cause considerable debate.

Lewis's final recommendation was that work should continue on the organically cooled reactor but that, given the amount of design work it still required, priority should be given instead to the fog-cooled and boiling-water reactors. Development of the heavy-water-moderated reactors would also receive attention; he suggested that Ontario Hydro should be approached concerning construction of more plants.[15]

At the next meeting this order of development was again called into question. Les Haywood argued in a report that there were significant differences in development costs between the different systems. On the basis of the papers presented, the committee decided to rank the heavy water system first, followed by the boiling-water and fog-cooled, and finally the organically cooled. However, after further comments had been made, Lewis admitted his concern that, by putting all their efforts towards the heavy-water model, they would eventually be "in a position where we could not compete with possible future developments."[16]

This fear led to a reordering of priorities, which was presented in a report of the committee. After outlining the different reactor types, the report recommended that both a 450-MWe and a 750-MWe heavy water reactor design should be incorporated into the Ontario Hydro system. However, the possible economic benefits of the organically cooled reactor required that its development continue. Although it could not receive highest priority, it was recommended that collaboration with other atomic agencies on these questions continue. Ongoing development of the fog-cooled and boiling-water reactors was recommended as well; once again, it would appear, the committee was unable to let go of any of the four different reactor types.[17]

Early in 1964 a debate erupted within the committee that would carry on for many months, eventually pitting engineers against scientists and affecting future models of CANDU. The controversy centred on the size of the pressure tubes. Up to this time they had been 3.25 inches in diameter but the question had been raised whether this should be increased to 4 inches. The deciding factor would be economics but it proved very difficult to determine which version would be cheaper. Figures in a report presented in January 1964 indicated that the wider the diameter of the tube, the lower the cost of uranium oxide fuel. But as there was some doubt about the accuracy of the numbers, the question was left for a later date.[18]

Les Haywood added his voice to the debate in a letter to Foster, Gray, and Lewis. He argued that the decision for larger tubes should be based on core size or clear economic advantage. In his view, the estimated economic gains to be made from the switch were not big enough. He pointed out that if a larger-size tube was chosen for future reactors, a great deal of work would be required to develop the larger fuel size. Further, they would have no reactor to test the larger fuel elements and problems might arise from the increase in fuel weight. In his opinion, it would be best to continue with the design of the CANDU-500 (i.e., a 500-MWe reactor) without too much further development.[19] Lewis agreed with Haywood that the advantages of the four-inch pressure tubes were not decisive. Although the fuelling costs would likely be slightly lower, the reactor would be larger in size and capital costs would go up. Lewis also feared that the development programs directed towards different reactor types (for example, the boiling-water reactor) would be displaced by the need for work on the pressure tubes.[20]

In late January, a meeting at Chalk River highlighted the differences between the two sides in the debate. The discussion, it would appear, was quite acrimonious; Haywood commented in a letter to Gray that it would have been "advantageous if a single balanced presentation of all the arguments could be made rather than having any single person or group identified with a particular stand. It is now not easy to be entirely objective in the consideration."[21] Haywood felt that enough information was available; it was simply a matter of clarifying it. He argued that the economic advantages of the larger tube diameter were not great enough. Although some development work had occurred, this was but a small step in the design, development, and testing necessary for an actual reactor design. Countering Gray's argument that the four-inch tubes would reduce heavy-water losses, Haywood stated that losses in the plant must be lower regardless and the difference of sixty tubes (which the larger tubes would bring) was insignificant. Others believed that the economic advantages of the larger tubes made the switch worthwhile. Foster argued that the price differences could amount to millions of dollars for each unit. Eventually, Gray asked the executive committee to decide the matter. In the end, Lewis and Haywood had to acquiesce to the engineers' wishes and the four-inch pressure tube became standard for the CANDU-500.

The argument over pressure-tube size was the central debate concerning the CANDU-500 reactor, but the larger question confronting the committee throughout the 1960s was which reactor types should be promoted and which should be dropped. This was not an easy decision to make. The first major report of the committee, issued in April 1963, did little to clarify the situation. Although it recommended that heavy-water reactor designs deserved priority, its authors could not let go of the other designs and suggested their continued development as well.

Clearly, PRDPEC members were loath to drop any design that showed promise. In part this was due to a fundamental fascination with the physics and engineering involved in the different reactor types. This was certainly the case with Lewis, who had long entertained himself by examining various possible reactors. But Lewis had another reason for maintaining study on all different designs. As noted in the minutes for a PRDPEC meeting early in 1964, he believed that "for the next fifteen years the export market for Canadian reactors should be larger than the domestic market. His impression is that foreign customers want a nuclear boiler with standardized design and components as much as possible and arrangements to permit the rest of the plant to be designed locally."[22] This impression contributed to his desire to leave the CANDU pressure tubes unchanged, and his belief in the future importance of export markets led Lewis to work towards maintaining development on all the different reactor designs.

The committee was not unaware of developments south of the border and the possible effects they might have on a future Canadian export market. The continuing interest of the Americans in light water reactors encouraged the Canadians to carry on with their work in this area. In fact it appears to have been a fear of being the only country focusing on heavy water reactors that motivated the committee to continue studying the four different reactor types even though only the heavy water CANDU variety was being built on a large scale.[23]

Lewis continued to push for power reactor development in order to stay competitive with other countries. Not everyone agreed with this motive. John Foster stated that "the major incentive for the development program has been to demonstrate the technical and economic feasibility of using our uranium resources for power production." Although Lewis would likely have agreed with that statement, Foster went on to dismiss the export market as a reason for future development. In his opinion, the export market would remain much less important than the domestic market. What was vital was proving that they had developed a "good and competitive product at home."[24]

These questions tied in directly with the future of AECL. By 1966 there was a sense that power reactor development could no longer be the central theme of the company and a new one would have to be found. In part this void could be filled by ING, which was then at the height of its development, but it was felt that other new ideas were required as well. Lewis argued that a number of projects needed to be developed simultaneously. As he put it, AECL scientists and engineers "must look for new ideas continuously, not merely when a current line of development seems to be terminating."[25]

PRDPEC began to meet less frequently in the late 1960s. By this time it had, in many respects, served its purpose. The CANDU reactor type had been developed successfully and was being constructed on a large scale at Pickering. The other reactor types, except the boiling light water, had not advanced to the prototype stage but their investigation helped develop the expertise of scientists at Chalk River. In the years before he retired, however, Lewis would use PRDPEC as a venue to present a final reactor type – the Valubreeder.

Lewis had long been interested in the use of thorium as a fuel. The fissile isotope of uranium, uranium-233, can be formed from thorium in much the same way that plutonium can be made from natural uranium. Since thorium is an abundant element, it was seen as an important potential source of energy when there appeared to be a serious shortage of uranium. Lewis remained interested in the potential of the thorium fuel cycle even after it became clear that there would be no uranium shortage and he wrote a number of papers on the subject in the late 1950s and early 1960s.[26]

For Lewis, the advantage of thorium (other than its abundance) lay in the fact that the fission properties of U-233 are superior to those of plutonium at the low neutron energies characteristic of CANDU and other thermal reactors. Lewis suggested using a fuel consisting of U-235 and thorium and continuing irradiation without reprocessing "to a burn-up of at least 1.5 fissions per initial U-235 atom supplied." Lewis argued that a fuel cycle based on this could be competitive.[27]

During the mid-1960s, when the battle for ING was at its height, Lewis does not appear to have had time to develop these ideas at great length. But in the spring of 1968, as it became increasingly clear that ING had little chance of receiving the necessary federal funding, Lewis wrote a paper detailing his conception of the "Valubreeder" reactor.

The Valubreeder was not a closely defined system and in fact covered a range of possibilities. As Lewis explained in a 1972 paper, "the essence of the valubreeder is to make the fuel supply predominantly natural uranium, together with a smaller amount of plain thorium and as a third component a small amount of the cheapest available fissile material."[28] With this setup, Lewis reported that "about 40% of the power comes from the fission of U233 bred into the thorium."[29] Lewis explained that the Valubreeder fuel cycle could be used in any of the CANDU reactors but added that the full advantages would only be gained when neutron economy was good. If a suitable reactor were used, however, the fuel cycle of the Valubreeder could be competitive over a long time. This made the reactor attractive for large-scale power complexes such as those needed for water desalination or agro-industrial processes.

In October 1968 Lewis presented his paper on the Valubreeder to PRDPEC. He expanded on the idea in some additional notes in which he maintained that the best reactor design for the Valubreeder would be one using organic coolant. There was some discussion at the meeting about Lewis's proposal. Les Haywood had difficulties with his claims concerning fuel savings. In his opinion, the savings would more likely come in a decreased capital cost in constructing the Valubreeder.[30]

Lewis raised the Valubreeder concept at another meeting in early January 1969 and again in July, but other committee members do not appear to have shown much interest. Perhaps in response to this, and in an attempt to gain support for the project, Lewis presented his ideas on the new reactor type to the board of directors in August 1969. Lewis outlined his views on the future of AECL's nuclear power program and argued that the large 1,500-MWe power station should be an organically cooled reactor using the Valubreeder fuel cycle. He noted that the use of natural uranium with a slight enrichment would continue to make the reactors attractive to customers in the future.

A number of objections were raised. Les Haywood pointed out that the engineers still remained to be convinced about Lewis's arguments. Furthermore, he believed that the question of organic cooling and the Valubreeder cycle should be considered separately. Another board member argued that they should continue to focus on the present power program. Unless that could be proven to be operating economically, other substantial changes in Canada's nuclear power program might be necessary. The minutes indicate that, despite Lewis's best attempt to convince board members of the merits of his proposal, for the time being no action would be taken to pursue it.[31]

In fact, the matter does not appear to have been raised with either the board or the executive committee for the next three years. Lewis continued to write on the subject but for the time being, at least, the idea was dormant. Even if Lewis was able to convince his colleagues that his ideas were physically viable, the problem of financing these huge projects still remained. In January 1973 Lewis presented to the committee a paper entitled "Canada's Opportunity," outlining his ideas on how to raise capital for the construction of future nuclear power plants in Canada. Noting that the energy crisis in the United States would lead to an increase in the price of oil, Lewis suggested that Canada impose an export tax on oil exported to the United States. The tax would be collected by Alberta, which in turn would act as banker. In its new-found status, Alberta could lend money at a low interest rate for the construction of nuclear power plants. These plants could be used to make synthetic fuels that would be needed when reserves were exhausted. Lewis believed this was a worthy idea and that AECL should present it to the government.[32]

The suggestion was poorly received. Lorne Gray bluntly stated that he could not accept Lewis's figures and that he would not identify

AECL with the proposal. Another director, David Golden, declared that the scheme was impractical. It required that either Alberta had to agree to less than the market price or the United States would have to pay more. Neither option was likely to occur.[33]

Lewis tried again a month later in front of the board of directors. The reaction was uniformly negative. A number of directors pointed out that the scheme had no chance of acceptance; as one director queried, why should the Alberta petroleum industry agree to a financial loss on the sale of its resource? Lewis also reported on a proposal he had made to Ontario Hydro. He had suggested that they raise electricity rates instead of borrowing capital funds. This would allow for greater investment in nuclear-plant installation and developmental work. But, as the minutes succinctly record, Ontario Hydro had shown no disposition to change its normal financing practices.[34]

As Lewis ventured further into the economic realm, his colleagues became less and less confident of the worthiness of his views. At the same board meeting, Lewis voiced his concerns over what he termed the deteriorating morale at Whiteshell resulting from the cancellation of the organically cooled reactor program. But Lorne Gray was not even able to concur with this assessment. Near the end of the meeting the notes record "general disagreement" with the proposal that Dr Lewis be called upon to act as a consultant for AECL at international meetings. Clearly Lewis's authority and position within the company were diminishing rapidly.[35]

Lewis tried to fight back at a subsequent meeting of the executive committee. It was reported that in a letter to Gray, Lewis indicated his refusal to accept the board's decision on the OCR program and questioned certain "allegations" made at the board meeting of 27 February. But again, the committee was united in its stand against Lewis and asked the president to write to him expressing their concern over the "damage that could be done to the AECL programme in an immediate sense and to Dr Lewis's personal credibility were he to attempt to force a public confrontation with the Committee and Board."[36] It is difficult to understand Lewis's loyalty to the unpopular Valubreeder proposal. Perhaps, in the wake of ING's cancellation, he sensed his declining influence in the company and responded by attempting to leave a final mark. His ideas, however, were clearly out of step with those of AECL's board of directors. And even if the reactor design was feasible, the questionable methods he suggested for financing its construction indicated, in this instance, a complete lack of political judgment.

Despite this rejection by his colleagues, Lewis defended the Valubreeder concept long into his retirement. In lectures given at universities across the country, he continued to promote nuclear power in general and the CANDU organically cooled, thorium- fuelled reactor in particular. Pointing to the success of the Pickering reactors as proof that nuclear power had to be developed, Lewis claimed that, with the different fuel and coolant of the Valubreeder, they could improve on the Pickering output by forty percent. Lewis did not limit his audience to universities; in 1977 he presented his views to the Royal Commission on Electric Power Planning.

Lewis's efforts did not lead to the construction of his suggested type of power station. The CANDU reactor used at Pickering functioned effectively and thus there was little incentive to spend the time or money required to develop Lewis's reactor, which he himself believed was essential for Canada's long-term energy needs. But for the time frame under consideration by the utilities, the power reactor already developed was sufficient.

Lewis's final decade with Atomic Energy of Canada Limited was in many respects very successful. He worked intimately with the scientists and engineers in the development of the CANDU reactor. His importance in this process is stressed by those who were involved in it with him. But Lewis also experienced a number of setbacks during this decade. In 1963, a company reorganization changed his title from Vice President (Research and Development) to Vice President (Science) and this was accompanied by a diminution of his responsibilities in the applied research sector. Four years later, a similar reshuffling left Lewis with just the programmatic direction of the basic science group. Finally, in 1969, Lewis became part of head-office staff and, although still located at Chalk River, had no remaining line responsibilities.

In part, Lewis's diminishing responsibilities at the plant must be attributed to a need to promote a new generation of younger scientists and administrators. But his failure to win government permission to proceed with the Intense Neutron Generator, followed by his unpopular adherence to the Valubreeder concept, must have contributed to a loss of power within the company. Ironically, the very characteristics that resulted in Lewis's strong influence in the power program at Chalk River – the depth of his knowledge, his tenacity and drive – ultimately led to a decline in his overall influence with the company in the final years before he retired.

CHAPTER NINE

Retirement

On 24 June 1973, Lewis turned sixty-five and retired from Atomic Energy of Canada Limited. It is not clear whether he had any plans for his retirement, but if so, they were put aside when he was offered a position at Queen's University in Kingston, Ontario. Alec Stewart, head of the physics department at the time, recalls being asked to a meeting with Queen's principal John Deutsch. Deutsch had heard that Lewis had retired and suggested to Stewart that he be invited to join the university's physics department. Stewart enthusiastically endorsed this idea and negotiations between Queen's and Lewis ensued.[1]

Lewis gladly accepted the offer for a new appointment. To indicate his position of honour at the university, he was given the title of Distinguished Professor of Science. For a person who would undoubtedly keep working long into his retirement, the situation was perfect. Lewis had an office at the department but was assigned no specific duties. He used it as a base from which to maintain his correspondence with people around the world, write lectures, and stay involved with a number of associations. Lewis generally spent the academic year at Queen's, returning to live in Deep River during the summer. He travelled a great deal to conferences around the world, always propounding his views on the need for nuclear power. For Lewis, the arrangement with Queen's was ideal.

In the years before his retirement and for many years afterwards, Lewis gave lectures defending the benefits of nuclear power. It was at this time that growing public awareness about the drawbacks of nuclear power began to lay the foundations of the antinuclear movement. From his lectures it would appear that Lewis was completely unable to understand this viewpoint. For him, any shortcomings of nuclear power were far outweighed by its advantages.

Until the 1970s, nuclear power had largely enjoyed the support of the general public in North America. Fear of nuclear war had waxed and waned in the quarter-century since the explosion of the first atomic bomb, but nuclear power had always been regarded as the positive side to the nuclear threat. This image had been introduced when the first United Nations conference on peaceful uses of atomic energy convened in 1955. During the 1960s, it was reinforced by the construction of nuclear power stations. It seemed that nuclear power promised a world transformed by nuclear energy.[2]

It was this viewpoint that W.B. Lewis expounded in lectures. He argued that mankind had a responsibility to learn how to control the abundant supplies of energy within its grasp. Only in this way could it hope to raise the standard of living and improve "the quality of life" of millions of people throughout the world.[3] Lewis did not discount hazards in the development and use of nuclear power, but in his view, these were necessary problems typical of every form of energy. A nuclear waste disposal area was analogous to a garbage dump. If properly planned and maintained, it need cause no threat to people's health. He dismissed the fear that background radiation from nuclear power plants might be detrimental to people's health. In fact, Lewis argued, it was possible that exposure to a low level of radiation might be beneficial, in much the same way that slight exposure to the sun's rays is healthy though overexposure can result in skin cancer. Lewis appeared to get carried away with the possible benefits of background radiation, pointing out that Indians living in the area of the monazite sands in Kerala are exposed to twenty times the typical background radiation of Bombay, and yet there is no rise in the incidence of cancer. He failed to suggest that the inhabitants might have adapted over thousands of years of living in that environment.[4]

In this and other articles, Lewis was reacting to the increasingly negative media publicity nuclear power was receiving. The media, Lewis believed, intent on "pulling scientists off their pedestals" and using scare tactics on the public, were not giving a fair representation of nuclear power. By focusing solely upon its negative aspects, journalists ignored the many helpful uses of nuclear power and radiation. Having already discussed the benefits of abundant, cheap power, Lewis outlined a few of the medical uses of radiation. Its use in agriculture and radio-biology had also proved widespread and beneficial – aspects of nuclear science, Lewis believed, that were generally ignored by the media. In the same vein, Lewis abhorred the idea of nuclear warfare, but added that the best way to avoid war was to develop nuclear power so that energy would be bountiful.

Lewis's arguments in these papers are mixed together in a manner that obscures any coherent viewpoint. Although he has clear ideas on these subjects, his inability to see or understand the other side of the argument makes his points less convincing. For Lewis, the benefits of nuclear power were so overwhelming that detractors did not deserve serious consideration. Perhaps he also underestimated the growing power of the antinuclear movement, which would increase steadily throughout the 1970s.

Lewis's arguments in favour of nuclear power reflected his positive outlook on the world. In his view, the development of nuclear power had offered mankind a virtually endless supply of energy that could be used to solve global problems. But while scientists had provided the possibility of limitless power, it was up to society to muster the will to develop it. It was this lack of will that Lewis found so frustrating and incomprehensible. In a lecture given in 1974 during the oil crisis, Lewis pointed out that there were no technical impediments to massive development of nuclear power plants; all that was needed was cooperation and trust among men.[5]

Lewis was correct in emphasizing the enormous contribution nuclear power could make but he was unable to comprehend the economic and, as the antinuclear movement became more vocal, political cost of its development. He continued to repeat his economic argument that large nuclear power stations should be paid for today by raising the cost of hydroelectric power, and perhaps also taxing fossil fuels, without recognizing how disastrous the political consequences would be. But if his arguments were sometimes faulty, his intentions were good. Lewis's optimistic outlook and hope for the future of mankind made him impatient with those who only saw disaster lying ahead. The "prophets of gloom" in the media had the wrong attitude towards the future. Lewis believed that more faith in mankind was all that was needed to achieve whatever goals were desired.[6]

Because of his enthusiasm for massive development of nuclear power, Lewis was doubtful about the need for fusion power. He held out the possibility of fusion reactors as a distant future goal and strongly believed in Homi Bhabha's 1955 prediction that the development of fusion would one day lead to unlimited quantities of power.[7] But the more immediate and realistic goal, he argued, should be nuclear power, not the least because of the innumerable technical problems involved in controlling fusion energy.

During his retirement Lewis had the opportunity to read and speak on questions concerning the future of man. Although he seldom touched on the controversial questions of nuclear weapons, his brief forays into this realm again emphasize his positive outlook. He drew from various authors in arguing that, as mankind developed, it inevitably moved towards greater peace and harmony. The chief mechanism of human evolution, he noted, was learning from the thoughts and experiences of others. Lewis believed that mankind had learned from its past experiences with war and was gradually moving towards a united and peaceful world. This positive perspective on man's future was linked, it would seem, with Lewis's Christian beliefs.

At the end of his career Lewis also briefly considered philosophical questions about science. He was particularly impressed with the ideas of Michael Polanyi, a Hungarian-British chemist and philosopher who wrote on how science and scientists functioned. Polanyi found a "republic of science" operating that was "a highly simplified example of a free society." According to Polanyi, to become a scientist you have to commit to the scientific system and the scientific world view. Scientists within this "republic of science" then work together to solve the problems presented by nature. When judging scientific work, Polanyi asserted, scientists must judge among themselves. The decision cannot come from a higher authority.[8]

In a paper also entitled "The Republic of Science," Lewis highlighted certain of Polanyi's ideas, which he felt were relevant for scientists at Chalk River. In particular, he quoted Polanyi's arguments that scientists must judge each other's work among themselves without looking to a higher authority. But the question how science would be directed to public welfare remained. Lewis argued this could be accomplished by impressing upon each scientist the importance of helping society. Then the "republic of science" would respond. Lewis rejected the idea that the government (a higher authority) should set out a national science program. In his view, if the political goals or policy were stated clearly, the republic of science would respond and answers to the problems would soon be forthcoming.

Polanyi's arguments struck a deep chord with Lewis. Even though a great deal of the scientific work he had been involved in throughout his career had been aimed towards a specific goal, Lewis had a strong belief in the importance of fundamental research. Lewis's support of cosmic ray research and the accelerator group are but two examples of the importance he placed on this type of research. For Lewis these pursuits were essential for Chalk River's continued support of applied research. The scientists in different areas overlapped, helped each other, and provided the intellectual atmosphere necessary for discovery. This same view was espoused by Polanyi, who argued that "self-coordination of independent initiatives leads to a joint result which is unpremeditated by any of those who bring it about."[9]

As he grew older, Lewis became less tolerant of those who disagreed with him. Nuclear power, he believed, offered so much promise and opportunity for the future of mankind that it was almost incomprehensible that anyone could argue against it. Part of the problem, perhaps, was that Lewis had spent very little time in examining the issue of radioactive waste. The question how ultimately to dispose of the radioactive fuel rods removed from reactors has haunted the nuclear power program from the beginning, but Lewis refused to admit that it was a real problem. In a couple of papers, he outlined a disposal method that involved encasing radioactive wastes in glass. But again, his presentation indicates a lack of understanding of public attitudes. In 1977 he published a paper recommending glass encasement and then storage in ground water. Although he emphasized that the water would be monitored to ensure that acceptable levels of radioactivity were maintained, it is hard to believe that Lewis did not foresee the public-relations problems inherent in such a suggestion. As his frustration grew, so did his arrogance; "people who see a problem storing radioactive wastes safely," he once declared, "are too ignorant to be trusted with the job."[10]

Such intolerance illustrates one of Lewis's significant failings. When convinced of the rectitude of his position through scientific and intellectual arguments, he simply could not understand the positions of people who did not agree with him. He had had this tendency throughout his career but it became exaggerated only as the tide turned against nuclear energy in the 1970s. Lewis responded with scientific arguments and criticisms of the "nay sayers" in the media but with minimal effect. Nuclear power did not experience a renaissance in popularity in his lifetime; in fact, the Three Mile Island accident in 1979 served to reactivate the antinuclear movement at a grassroots level.[11]

Lewis's attempts to promote the development of nuclear power had little effect on protest groups in society but he continued to be honoured by his peers. Throughout his career, Lewis had received many prestigious awards. His scientific accomplishments at the Cavendish and at TRE during the war were recognized through his election as a fellow of the Royal Society in 1945. For his role in the development of radar he was named a Commander of the Order of the British Empire in 1946 and awarded the American Medal of Freedom with Silver Palms a year later. Lewis was involved in the American Nuclear Society from its inception and was named president in 1961, the only Canadian to be so honoured. In 1966, the Canadian government acknowledged Lewis's important contributions to the nuclear power program by giving him the first Outstanding Achievement Award of the Public Service of Canada. This was followed in 1967 by the Atoms for Peace Award, which he received along with Bertrand Goldschmidt and I.I. Rabi. Other awards and honorary degrees peppered Lewis's career, but perhaps the greatest accolade bestowed upon him was the Fermi award in 1982. Among the list of previous recipients were such scientific luminaries as John von Neumann, E.O. Lawrence, Hans Bethe, J. Robert Oppenheimer, and Otto Hahn. It was an honour for Lewis and for Canada to receive this award.

Sadly, Lewis was unable to travel to Washington for the award presentation; instead a small ceremony was held at Chalk River. Lewis's inability to travel was the result of a car accident a year earlier while driving back to Kingston from Deep River. At first it seemed that he had suffered little damage, but a few days later a cracked cervical vertebra was discovered. Wearing a brace, Lewis continued to function both at Queen's and in Deep River. It soon became clear that something much more fundamental was wrong, however. Shortly after he received the Fermi award in early 1982, Lewis was diagnosed as having Alzheimer's disease. For the remaining five years of his life, he was cared for at the Deep River hospital and visited daily by his sister. He died there on 10 January 1987.[12]

Conclusion

In 1973, the year Lewis retired from AECL, Jacob Bronowski published a book entitled *The Ascent of Man*. Lewis was very impressed by this book and recommended it to many of the people with whom he corresponded.[1] The book traces the cultural evolution of mankind, describing and interweaving major developments over the course of human history. It examines how each scientific achievement fits into a chain of knowledge that advances the growth of humanity. For Lewis, Bronowski's book voiced his personal belief in the benefits of science and the positive effects it has had on the upward struggle of humankind.

Lewis believed in science. He was convinced that scientific knowledge, if properly utilized, would materially benefit people's lives. His work first on radar and subsequently on nuclear energy were Lewis's contribution to society. These were further small steps in the ascent of man.

Lewis was steeped in science all his life. He broke with the engineering tradition within his family to pursue experimental physics but maintained a practical streak in his approach. His years at the Cavendish embedded within him a strong belief in the importance of basic research. Throughout his career, he maintained this allegiance, which ensured that it became a fundamental part of the Chalk River mandate.

The Second World War brought about a crucial change in Lewis's career path, as his years as a university scientist performing research and giving lectures came to an end. Instead, during the war, he helped in the scientific direction of a radar research establishment. The scientific expertise necessary for the successful development of radar became vital to the war effort. The scientists themselves in turn were essential and recognized their indispensable role. To an unprecedented extent they helped to shape policy decisions. Lewis participated in this process and became convinced that the scientist's policy-making role was legitimate – a conviction he would maintain throughout his career. What held true in wartime, however, did not necessarily carry over to times of peace. The influence of scientists on policy during peacetime depended on their recognition that political and economic limits exist. At times during the latter part of his career, Lewis seemed to have lost sight of this.

Lewis thrived during the war when working, together with other scientists, on an urgent mission closely linked to his country's national security. To tackle the task at hand, he made the most of his talent for assimilating large amounts of information and directing it towards a goal. This ability would be crucial in his leadership of Canada's nuclear-power program in the postwar period.

In his position as scientific director at Chalk River, Lewis began to pursue a new objective. He recognized the potential of nuclear power and persuaded AECL's board of directors to accept his vision of a Canadian program. Developing that program was an enormous undertaking that demonstrated Lewis's tenacity in his singleminded pursuit of a goal. Once he was convinced that nuclear power was a worthy objective, it not only became the mission of Chalk River but also Lewis's own personal mission. Lewis, it can be said, "preached" the advantages of nuclear power from the early 1950s until his final illness.

Lewis's confidence in the benefits of nuclear power for this country and in Canada's ability to build a safe, efficient, and economic system were well suited to the mood of the country in the prosperous postwar years. Canada's role in the allied atomic energy effort during the war ensured its membership in the atomic club. Lewis acted as Canada's representative on the scientific advisory committees of both the United Nations and the International Atomic Energy Agency. Participation in international scientific activities and collaborative efforts with underdeveloped countries were in tune with Canadian foreign policy of this period. This linkage between different policy areas provided support for Canada's growing nuclear program.

Public backing of nuclear power was strong in North America throughout the 1950s. During the 1960s and 1970s, however, the growth of the environmental movement led to the formation of groups critical of nuclear power. This was part of a wider suspicion of science in general. Where once science had seemed able to provide solutions to the world's problems, it now seemed to be the cause of many of them. Lewis reached the height of his career when support for nuclear power was at its zenith; as the popular allure began to fade, both Lewis and the laboratories he directed experienced significant setbacks.

To a certain extent Lewis contributed to Chalk River's diminished support. By nature a very private man, Lewis was wholly dedicated to his work and willing to immerse himself completely in it. In the early days of Chalk River, this exactly matched the mood of the laboratories. In their isolated location the scientists worked intensely in a new area of physics. Generous government support made the situation close to ideal.

But this state of affairs could not last forever. By the 1960s, research centres at universities were gaining strength and questioning why government science in general, and Chalk River in particular, were receiving such a large proportion of the country's science budget at a time when that budget was shrinking. Lewis did not respond well to this change in mood. His actions during the debate surrounding the Intense Neutron Generator showed that he was increasingly unable to understand the viewpoint of those who opposed the proposed facility. The isolated nature of Chalk River combined with Lewis's inherent character traits to make him impervious to the new rules of the game.

These drawbacks do not detract from Lewis's strong contributions to science in Canada. His commitment to nuclear power existed on a number of levels. He was fascinated by it scientifically and as a technological feat. He also viewed it as crucial for Canada's growth. Lewis believed that for a country to grow, it needed power; in Canada, that power would be supplied by nuclear energy. On another level, his enthusiasm for nuclear power tied in with his humanitarian beliefs. Canada could share its knowledge and expertise to help underdeveloped countries.

The most visible product of these efforts was the CANDU reactor. Lewis played a central role in the development of this distinctive Canadian reactor design in the 1950s. To be sure, other scientists and engineers made significant contributions, but as scientific leader at Chalk River, the final decisions fell to Lewis. His choice of heavy water and natural uranium for the new design reflected both a desire to build upon Chalk River's strengths and a tested conviction that these were optimal components for a power reactor design. Lewis's natural scientific curiosity led him to continue his examinations of different reactor types, but he retained an unshakeable belief in the CANDU.

In many respects, this belief was justified. CANDU has proven to be an efficient reactor whose performance, when compared to that of other power reactors around the world, is consistently rated in the top ten. Nearly half of the electricity produced in Ontario is nuclear-generated at rates competitive with fossil-fuel-fired plants. More generally, the scientific infrastructure that spawned CANDU has provided the basis for a fruitful research program at both Chalk River and Whiteshell.

At the same time there are, without doubt, problems with Canada's reactor design. Recent difficulties with the pressure tubes have proven costly both in dollar terms and in the resulting increase of public concern over nuclear power. That concern is compounded by the persistent failure to find a means of long-term disposal of radioactive waste produced in reactors. Furthermore, the costs of waste disposal and of decommissioning old reactors are still not fully known.

Some of these problems are a direct result of the CANDU design, but most apply to nuclear power reactors around the world. The spectacular predictions of the wonders of nuclear power voiced by Lewis and his counterparts in other countries during the 1950s and 1960s simply have not come to pass. As a result, the CANDU reactor has been a mixed blessing for Canada. It brought the prestige of being part of the atomic club, and recrimination when the promise of nuclear power fell far short of expectations. Although Lewis must be criticized for his excessive faith in that promise, he deserves praise for the vision and determination that made CANDU possible.

At Chalk River, Lewis's most important legacy would be the strong tradition of basic research he enshrined there. His belief in the importance of fundamental research led him to initiate and push through plans for a second research reactor. Under his guidance, scientists at Chalk River achieved excellence in a number of fields. The laboratories have maintained those standards to the present. Lewis's influence at the facility was great and he was still regarded as a leader even into his retirement. In Deep River, where he continued to spend his summers, he maintained his position on the library board. Although not everyone agreed with Lewis's management practices at Chalk River and in the town, everyone agreed that his influence had been immense.

Lewis dedicated the better part of his career almost entirely to the pursuit of nuclear power. He immersed himself in his work and spent very little time on other activities. This complete concentration on work makes it difficult to get a true understanding of what drove him. It is possible, perhaps, to see a clue in some thoughts he put to paper as a young man. In documents Lewis saved from his Cambridge days, there are rough notes he prepared for a sermon that he may have been invited to give at his college chapel. In one section he stated the theme of his sermon as follows: "In order to live as distinct from drifting through life we need a purpose and a background. We can arrive at a purpose by reflection and study, we find out what kind of lives men have lived and our tremendously complicated sense of right and wrong, helps us to find a purpose for ourselves. Finding a background is not so simple. I suppose that all those here have adopted as the general background a sense of God having active relation with the world."[2]

Lewis's belief in God shaped the larger philosophical background of his life, but one can also see his belief in science and its benefits for mankind as a constant theme. This conviction provided him with the purpose necessary for his life to be worthwhile. During the war radar was his central mission. Later, nuclear power became his principal pursuit. In the end, Lewis had a fundamental effect on nuclear research in Canada. No one person had greater influence over the direction of Canada's nuclear power program. The tradition of excellence in nuclear research at Chalk River and the CANDU reactor design remain as Lewis's legacy to Canada.

Appendix

BIBLIOGRAPHY OF WILFRID BENNETT LEWIS

The following list of articles by W.B. Lewis is divided into three main sections: "United Kingdom," "Chalk River," and "Retirement." Articles in the Chalk River section are further divided into four sections: Director's Lectures (DL), Director's Memoranda (DM), Director's Reports (DR), and AECL reports. These can be obtained from either the library or the Scientific Documents Distribution Office at the Chalk River Nuclear Laboratories.

UNITED KINGDOM PUBLICATIONS

"Note on the Problem of Selectivity Without Reducing the Intensity of the Sidebands." *Experimental Wireless and Wireless Engineer* 6 (March 1929): 133–4.

"The Transmitting Station actually sends out Waves of one Definite Frequency but of Varying Amplitude." *Experimental Wireless and Wireless Engineer* 6 (May 1929): 261.

"The Apparent Demodulation of a Weak Station by a Stronger One." *Experimental Wireless and Wireless Engineer* 8 (October 1931): 538–40.

"Analysis of the long-range alpha-particles from radium C." Written with Lord Rutherford and F.A.B. Ward. *Proceedings of the Royal Society of London* A 131 (1931): 684.

"Analysis of the alpha-particles emitted from thorium-C and actinium C." Written with Lord Rutherford and C.E. Wynn-Williams. *Proceedings of the Royal Society of London* A 133 (1931): 351.

"The Detector." *Experimental Wireless and Wireless Engineer* 9 (September 1932): 487–99.

"Demodulation." *Experimental Wireless and Wireless Engineer* 9 (November 1932): 629–30.

"The range of the alpha-particles from the radioactive emanations and 'A' products and from polonium." Written with C.E. Wynn-Williams. *Proceedings of the Royal Society of London* A 136 (1932): 349.

"Analysis of alpha-rays by an annular magnetic field." Written with Lord Rutherford, C.E. Wynn-Williams, and B.V. Bowden. *Proceedings of the Royal Society of London* A 139 (1933): 617.

"Analysis of the long range alpha-particles from radium C′ by the magnetic focussing method." Written with Lord Rutherford and B.V. Bowden. *Proceedings of the Royal Society of London* A 142 (1933): 347.

"An analysis of the fine structure of the alpha-particle groups from thorium C and of the long range groups from thorium C." Written with B.V. Bowden. *Proceedings of the Royal Society of London* A 145 (1934): 235.

"Improvements to the 'scale of two' thyratron counter." *Proceedings of the Cambridge Philosophical Society* 30 (1934): 543.

"Attempts to detect gamma-radiation excited by the impact of alpha-particles on heavy elements." Written with B.V. Bowden. *Philosophical Magazine*, 7th ser., 20 (1935): 294.

"Experiments with high velocity positive ions. V. Further experiments on the disintegration of boron." Written with J.D. Cockcroft. *Proceedings of the Royal Society of London* A 154 (1936): 246.

"Experiments with high velocity positive ions. VI. The disintegration of carbon, nitrogen and oxygen by deuterons." Written with J.D. Cockcroft. *Proceedings of the Royal Society of London* A 154 (1936): 261.

"A Portable Duplex Radio-Telephone." Written with C.J. Milner. *Experimental Wireless and Wireless Engineer* 13 (September 1936): 475–82.

"An attempt to produce artificial radioactivity by an electron beam, with some notes on the behaviour of newly made Geiger-Müller counters." Written with W.E. Burcham. *Proceedings of the Cambridge Philosophical Society* 32 (1936): 503.

"A repetition of the Bothe-Geiger experiment." Written with W.E. Burcham. *Proceedings of the Cambridge Philosophical Society* 32 (1936): 637.

"Transient Response." *Wireless World* 41 (1937): 250.

"Alpha-particles from the radioactive disintegration of a light element." Written with W.E. Burcham and W.Y. Chang. *Nature* 139 (1937): 24.

"A scale of two high speed counter using hard vacuum triodes." *Proceedings of the Cambridge Philosophical Society* 33 (1937): 549.

"The multi-electrode valve and its application to scientific instruments." *Journal of Scientific Instruments* 15 (1938): 353.

"On the production of radium E and polonium by deuteron bombardment of bismuth." Written with D.G. Hurst and R. Latham. *Proceedings of the Royal Society of London* A 174 (1940): 126.

Electrical Counting. Cambridge: Cambridge University Press, 1942.

CHALK RIVER PUBLICATIONS

Director's Lectures

DL-1 "Resistors in Electronic Apparatus." 25 March 1948.
DL-2 "Some History on the Development of Radiation Counting Techniques." 2 June 1949.
DL-3 "Measurement of Energy of Charged Particles." 12 July 1949.
DL-4 "Radioactive Contamination." 9 August 1949.
DL-5 "Whither Atomic Energy Research?" 20 October 1949.
DL-6 "The Gleam in the Eye of the Atomic Scientist." 15 November 1949.
DL-7 "Introductory Remarks on Philosophy of Design." 22 March 1950.
DL-8 "100 Years of British Scientific Development." 12 March 1951.
DL-9 "The NRX Pile at Chalk River in Operation for Physics Research." June 1951.
DL-10 "Chalk River Research Review and Prospect." 18 February 1953.
DL-11 "The Goose that Lays the Golden Eggs." 13 March 1953.
DL-12 "Atomic Power Progress and Prospects." September 1953.
DL-13 "Atomic Energy Outlook for the Future." 7 December 1953.
DL-14 "Abundant Power to be Won by Work." 11 February 1954.
DL-15 "Canadian Power Development – Future Possibilities." 1954.
DL-16 "Economic Aspects of Nuclear Power." 1955.
DL-17 "Possibilities of Generating Atomic Electric Power on a Competitive Basis." 1955.
DL-18 "Atomic Energy A Source of Future Power – Its Impact on Industry." 1955.
DL-19 "Economic Power Fuelling Without U-235 Enrichment." 12 December 1955.
DL-20 "Reactor Design and Technology." August 1955.
DL-21 "Atomic Energy 1955." 10 November 1955.
DL-22 "The Canadian Research Reactors and Their Uses." July 1956.
DL-23 "Atomic Energy a Source of New Power and New Knowledge." 4 April 1956.
DL-24 "Why Heavy Water?" 5 June 1956.
DL-25 "The Heavy Water Reactor for Power." May 1956.
DL-26 "Canadian Experiments Aim at Economic Nuclear Power." June 1956.
DL-27 "Energy Prospects." 14 June 1956.
DL-28 "Scientific Prospects in Atomic Energy." 13 December 1956.
DL-28A "Scientific Prospects in Atomic Energy." 1957.
DL-29 "Natural Uranium – Heavy Water Reactors for Low Cost Power." 15 March 1957.

DL-30 "The Efficient Use of Manpower." June 1957.
DL-31 "Natural Uranium and Enriched Uranium, Technical and Economic Aspects." 1957.
DL-32 "Reactor Fuel Reprocessing – The Major Consideration in Canada That May Be Different From the United States." 28 October 1957.
DL-33 "Uranium Oxide Fuel of Low Cost." 17 March 1958.
DL-34 "Uranium Dioxide." 8 December 1958.
DL-35 "Experiences With Loops in the NRX Reactor." June 1959.
DL-36 "Experiences in the Irradiation of UO2 Fuel." October 1959.
DL-37 "Planning Canada's Nuclear Programme." September 1959.
DL-38 "Reality – Nuclear Power Plants in Canada." November 1959.
DL-39 "Neutron Economy." 19 January 1960.
DL-40 "Nuclear Reactors." 19 February 1960.
DL-41 "Dreams and Gleams in Applied Research." 30 June 1960.
DL-42 "Designing Heavy Water Reactors for Neutron Economy and Thermal Efficiency." January 1961.
DL-43 "Pressure-Tube Heavy-Water-Cooled Reactors." October 1960.
DL-44 "The Canadian Nuclear Power Program." April 1961.
DL-45 "Behaviour of Fission Gases in UO2 Fuel." 1961.
DL-46 "Man and Machine in Reactor Control." February 1962.
DL-47 "A Canadian View of the United States Civilian Power Reactor Program." 26 February 1962.
DL-48 "Operating Experience with Heavy Water Nuclear Power Reactors." June 1962.
DL-49 "Giant Power: Planning the Contribution and Growth of Nuclear Power." April 1962.
DL-50 "Nuclear Electric Power – One of the Giants." 19 June 1962.
DL-51 "Criteria for the Selection of Materials for Water-Cooled Reactors, with Comments on D_2O Reactors." October 1962.
DL-52 "An Assessment of Plutonium Use." Written with O.J.C. Runnalls. September 1962.
DL-53 "The Growth and Yield of Atomic Energy Research." November 1962.
DL-54 "Prospects in Atomic Energy Research." March 1963.
DL-55 "A Perspective on Direct Conversion." October 1963.
DL-56 "Tomorrow's Fission Reactor and Its Materials Demands." February 1964.
DL-57 "Lecture Report on Visit to the USSR." 1 July 1963.
DL-58 "Fuelling High Conversion Ratio Reactors." September 1963.
DL-59 "Experience with Canada's First Nuclear Power Station (NPD) and the Prospects for Heavy Water Power Reactors." 1964.
DL-60 "Choice and Change." 12 May 1964.
DL-61 "Nuclear Power Development in Canada and Other Countries." May 1964.

DL-62 "Some AECL Aims in Applied Research." 22 June 1964.
DL-63 "Canada's Position in the World as Seen at Geneva." October 1964.
DL-64 "Government Funded Research – Government and Industry." 2 November 1964.
DL-65 "Introduction to the Symposium on the Generation of Intense Neutron Fluxes." April 1965.
DL-66 "Some Problems of Interdependence in Metallurgical Research." May 1965.
DL-67 "Heavy Water Reactor Review and Prospect." June 1965.
DL-68 "Engineering for the Fission Gas in UO_2 Fuel." April 1966.
DL-69 "The Future of Nuclear Energy." 9 May 1966.
DL-70 "The Intense Neutron Generator Study." 30 May 1966.
DL-71 "Power Reactor Systems Research." October 1966.
DL-72 "The Intense Neutron Generator." 25 October 1966.
DL-73 "Achievements and Prospects of Heavy Water Reactors." 1966.
DL-74 "Canada's Role in Atomic Energy." February 1967.
DL-75 "Canada's Planned Advanced Reactor Concepts." April 1967.
DL-76 "Nuclear Energy in the Future." 15 March 1967.
DL-77 "Nuclear Power Development in Canada." March 1967.
DL-78 "Fuel and Finance Cycles for Nuclear Power in the Long Term." April 1967.
DL-79 "Prospective Competitive Cycles of Fuel, Power and Finance for Delivering Electric Power." 12 April 1967.
DL-80 "General Philosophy of Heavy Water Reactor Systems." N.d.
DL-81 "Economics of Nuclear Fuel Cycles." 22 September 1967.
DL-82 "Economic Relations Between Fast and Heavy Water Power Reactors." 28 September 1967.
DL-83 "Prospect for Heavy Water Reactors." December 1967.
DL-84 "Some Preliminary Thoughts on Ion Drag Accelerators." February 1968.
DL-85 "Economic Heavy Water Power Reactors." May 1968.
DL-86 "Future Factory Type Accelerators." May 1968.
DL-87 "A Look to the Future." 19 April 1968.
DL-88 "The Future of Nuclear Energy." 12 March 1968.
DL-89 "The Importance of Atomic Energy to World Development." 22 March 1968.
DL-90 "Accelerators for Intense Neutron Sources." March 1968.
DL-91 "Near Prospects in Nuclear Energy." 21 March 1968.
DL-92 "25 Years Since Decision." June 1968.
DL-93 "Review of Experience with Heavy-Water-Moderated Power Reactors." 1968.
DL-94 "The Intense Neutron Generator and Future Factory Type Ion Accelerators." 1968.

DL-95 "Ion Drag Acceleration." 28 November 1968.
DL-96 "Mutual Confidence the Key to Unlimited Energy." April 1969.
DL-97 "Perspective in Atomic Energy." 1 May 1969.
DL-98 "The Future of Heavy-Water Reactors." June 1969.
DL-99 "Science in AECL." 27 October 1969.
DL-100 "Thermal Pollution or Enrichment." N.d.
DL-101 "Canadian Operating Experience with Heavy Water Power Reactors." Written with J.S. Foster. April 1970.
DL-102 "Economics of Nuclear Power." 13 February 1970.
DL-103 "Canada, the Quality of Life and Nuclear Energy." 1970.
DL-104 "Brief Historical Introduction to Mercury Hazards." February 1971.
DL-105 "Recovery and Benefits from Radiation Damage." 25 January 1971.
DL-106 "Frontier Events in Electrical, Electronic and Radio Engineering from Picowatts to Terawatts in the Last 40 and the Next 60 Years." 19 May 1971.
DL-107 "Foreseen Development of CANDU Reactors to 2000 A.D." June 1971.
DL-108 "Nuclear Energy and Radiations from Rutherford at McGill to the Present." October 1971.
DL-109 "Nuclear Energy and the Quality of Life." N.d.
DL-110 "Advanced HWR Power Plants." August 1972.
DL-111 "Past and Future." 30 May 1972.
DL-112 "Choice of Nuclear Reactors for Power." 21 June 1972.
DL-113 "Meeting the World's Energy Needs." 6 July 1972.
DL-114 "The Significance of Thorium for Nuclear Energy." 1 December 1972.
DL-115 "Development Potential of CANDU Reactors." N.d.
DL-116 "CNA Luncheon Address." 19 June 1973.

Director's Memoranda

DM-1 "Notes on a meeting to discuss plutonium monitoring." 12 November 1946.
DM-2 "Notes on a meeting to discuss contamination control." 30 November 1946.
DM-3 "Notes of a meeting to discuss factors affecting the choice of form of thorium in the NRX pile and future piles, including high temperature, power and breeder piles." 5 February 1947.
DM-4 "Notes on proposed investigation of the blistering phenomenon of uranium rods discussed at meeting." 14 April 1947.
DM-5 "Notes of a meeting to discuss standard sources and measurements." 15 April 1947.
DM-6 "Notes on meeting to discuss program for irradiation of uranium." 4 March 1948.
DM-7 "Notes on meeting held to discuss procedures for analytical work." 27 May 1949.

DM-8 "Notes on meeting held to discuss tritium preparation and uses." 26 June 1948.
DM-9 "Notes on a meeting with the instrumentation representatives of the USAEC held in New York." 2 December 1948.
DM-10 "Notes on discussions in the Physics Group at the Isotopes Conference held at the NRC." 7 December 1948.
DM-11 "Notes on a 2d General Meeting to discuss corrosion of the calandria." 4 August 1949.
DM-12 "Notes on a meeting held to discuss U-236." 21 November 1949.
DM-13 "Program for measuring low level radioactivity in water." 17 July 1950.
DM-14 "Notes of preliminary meeting to discuss implementation and program of pile instrumentation for NRU." 2 October 1950.
DM-15 No copy in file.
DM-16 No copy in file.
DM-17 "Minutes of a meeting to discuss policy for operations and development of the plutonium extraction plant. 21 December 1950.
DM-18 "Projects generating from the Atomic Energy Project." 19 April 1951.
DM-19 "Notes on preliminary meeting re Xenon-135." 24 August 1951.
DM-20 "Tritium formation in the NRX pile heavy water." 7 September 1951.
DM-21 "Minutes of meeting to discuss program of Chemical Engineering, Research Chemistry and Metallurgy Branches for Power Reactors." 5 February 1954.
DM-22 Number not allotted.
DM-23 "Small power reactor design for pressurized heavy water natural uranium 50 megawatt reactor." 6 April 1954.
DM-24 "Selling prices of plutonium and depleted uranium." 20 April 1954.
DM-25 "The proposed Canadian "Boiling Water" Power Reactor." 30 August 1954.
DM25A Table to accompany DM-25.
DM-26 "Comparison of Pippa II and heavy water recycling reactor for Canadian power program." 14 October 1954.
DM-27 "The importance of epsilon." 20 April 1955.
DM-28 "Development and preliminary design study programme for the two large power reactors." 21 April 1955.
DM-29 "Required bond strength for NRU rod sheathing. Minutes of a preliminary meeting to discuss the problem before setting a program." 25 April 1955.
DM-30 "Canadian Optimum Heavy Steam Reactor I." 6 July 1955.
DM-31 "Capture Cross-Section of PU-241." 18 July 1955.
DM-32 "Technical policy on fuel sheathing." 30 July 1955.
DM-33 "International aspects of atomic energy." 12 September 1955.
DM-34 "Fuel for water-cooled power reactors." 28 September 1955.
DM-35 "Fuel rod monitoring requirements of NPD." 3 October 1955.

DM-36 "Note on enrichment required in a reactor having any fuel distribution." N.d.

DM-37 "Possible use of ThO_2 as fuel in NRX." 18 April 1956.

DM-38 "Comment on study by Corbin Allardice, Adviser on Atomic Energy, International Bank for Reconstruction and Development, entitled 'Economic Nuclear Power Today: Where and under What Circumstances.'" 10 July 1956.

DM-39 "Burn-up attainable from plutonium recycling in thermal reactors." 22 October 1956.

DM-40 "Note on the importance and practicability of energy storage with nuclear power." 29 January 1957.

DM-41 "Electric power prospects in Canada." 28 January 1957.

DM-42 "Fuelling System for Natural Uranium Reactor with Long Rods for High Burn-up without Recycling." 1957.

DM-43 "Comparison of reactors for low cost power." 1 May 1957.

DM-44 "The Significance of Developing a High Performance Uranium Oxide Fuel." 1957.

DM-45 "Note on resonance escape probability calculated by Critoph's method of AECL #350." N.d.

DM-46 "Refined evaluation of the long irradiation of uranium fuel." 1 August 1957.

DM-47 "High Burn-up from Fixed Fuel." 1957.

DM-48 "Comment on American Standard's 'An Evaluation of Heavy Water Reactors for Power.'" 1957.

DM-49 "Economics for Ontario Power." 20 January 1958.

DM-50 "Controlled H-fusion." 29 January 1958.

DM-51 "A synoptic atomic energy programme for Canada." 1 April 1958.

DM-52 "Long Irradiation of Natural Uranium." 1958.

DM-53 "Note on the power reactor development programme." 5 August 1958.

DM-54 "Initial Fuelling of Power Reactors." 1958.

DM-55 "Carbon-14 in geophysics." 8 December 1958.

DM-56 "Minutes of meeting to discuss the application of recent technology to the development of CANDU." 6 February 1959.

DM-57 "Basic Consideration in the Design of the Full Scale Heavy water and Natural Uranium Power Reactors." 1959.

DM-58 "The Return of Escaped Fission Product Gases to UO2." 1960.

DM-60 "Nuclear power possibilities for Canada." 11 April 1960.

DM-61 "Estimated burn-up from CANDU." 7 September 1960.

DM-62 "Calculations on the long irradiation of uranium." N.d.

DM-63 "The practicability of 50% efficiency from a nuclear reactor steam cycle." N.d.

DM-64 "Optimizing Organic-Liquid-Cooled Heavy-Water Natural- Uranium Reactor Design for Shut-Down Refuelling." 1961.

DM-65 "Requirements for OTR and the development program." 10 October 1961.
DM-66 "Comment on 'Fuel Cycle Analysis for Successive Plutonium Recycle 1. Results for Five Reactor Concepts,'" by E.A. Eschbach et al." 24 April 1962.
DM-67 "Canada 1962." 23 July 1962.
DM-68 "Pressure of Fission Product Gases in Clad UO_2 Fuel." N.d.
DM-69 "Breeders Are Not Necessary – A Competing Other Way for Nuclear Electric Power." 1963.
DM-70 "Research needs for large-scale nuclear fission power." N.d.
DM-71 "Inspiration and aims for advanced projects in Canada." 24 October 1963.
DM-72 "How Much of the Rocks and Oceans for Power? Exploiting the Uranium-Thorium Fission Cycle." 1964.
DM-73 "Long term conclusions from the nuclear fuel inventory problem." N.d.
DM-74 "The thorium seed-blanket large power reactor (LPR) project." July 1964.
DM-75 "A thorium + U-235 challenger to CANDU-500." July 1964.
DM-76 "Proton accelerator for intense neutron generator." July 1964.
DM-77 "The aims and desirability of the continued growth of AECL." August 1964.
DM-78 "Proton accelerator for intense neutron generator." January 1965.
DM-79 "The significance of the diametral strain of pressure tube MK IV in U-2 look." March 1965.
DM-80 "Note on spatial reactivity vs. flow stability problem in CANDU-BLW reactors." April 1965.
DM-81 "Travelling wave proton accelerators and the staticelerator." August 1965.
DM-82 "Ruptured nuclear fuel location by gamma-ray survey of traps." August 1965.
DM-83 "The problem of meaningful definitions of uranium inventory for power reactor systems." N.d.
DM-84 "Neutron absorption in fission products." March 1966.
DM-85 "Canadian interest in developing the disc generator." April 1966.
DM-86 "Disc generator for high voltage D.C." August 1966.
DM-87 "Proposed experiments to test a new principle for a high energy accelerator." February 1967.
DM-88 "Design proposal for a magneto generator for megawatt power at high voltage exploiting varium ferrite magnets and a strong insulated shaft." 18 October 1966.
DM-89 "A perspective in Canada on expenditure for the intense neutron generator." November 1966.

DM-90 "Note on production of plutonium-238 in NRU and ING." February 1967.

DM-91 "A new policy for AECL (and other government laboratories)." N.d.

DM-92 "Preliminary comparison of ion drag accelerator proposals." January 1968.

DM-93 "Major new activities in science and technology." 12 February 1968.

DM-94 "The Super-Converter or Valubreeder. A Near-Breeder Uranium-Thorium Nuclear Fuel Cycle." 1968.

DM-95 "To maintain Canada's selected roles in atomic energy." 24 April 1968.

DM-96 "Prospects for 44% Cycle Efficiency from 400C Steam in Large Electric Generating Stations." 1969.

DM-97 "Immediate Relation of ING to Fast Breeder Reactor Programs." 1969.

DM-98 "The role of AECL in Canada's energy development." 10 November 1968.

DM-99 "Electron ring accelerator -1650B injector." December 1968.

DM-100 "Development for fast breeder reactors." January 1969.

DM-101 "The importance of fundamental research in AECL." 22 January 1969.

DM-102 "Preliminary basic consideration for the design of an organic liquid-cooled heavy-water-moderated 1500 MW(e) power reactor." 24 February 1969.

DM-103 "Adapted computer code estimates for a 1500 MW(e) organic liquid-cooled heavy-water-moderated power reactor." May 1969.

DM-104 "Interim note on a relation linking fast neutron radiation damage, sintering, stress-rupture, ternary creep, radiation-induced creep and gas absorption growth and shrinkage of voids, pores and gas bubbles in metals and other solids." 21 January 1971.

DM-105 "Comparison of Power Cost Projections." 15 June 1969.

DM-106 "Co-operative competition with fast breeder reactors." July 1969.

DM-107 "To L.R. Haywood: Effect of uranium price on valubreeder and natural uranium cycles." 21 July 1969.

DM-108 "AECL policy for the 1500 MW(e) reactor, presentation to the Board of Directors." 14 August 1969.

DM-109 "Reactor physics comparison of CANDU-OC-VB with SGHW." 20 November 1969.

DM-110 "Comment on 'Report on Controlled Fusion Research – Ottawa, June 1969' by the H.M. Skarsgard Sub-Committee of the National Research Council Associate Committee on Plasma Physics." 5 January 1970.

DM-111 "Tritium retention in thermonuclear fusion reactors." 28 August 1970.

DM-112 "Prospect of Very Low Cost Power from Ontario Hydro Based on Large Capacity Nuclear Generating Stations." 1970.

DM-113 "Separative work contribution to fuel cycle costs, related to Uranium-235 enrichment by centrifuge for the valubreeder." May 1970.

DM-114 "Cobalt-60 production and enrichment in NPD." March 1966.

DM-115 "Designing UO_2 fuel for negligible sheath strain." N.d.

DM-116 "Power reactor development (to all members of PRDPEC)." 24 August 1970.

DM-117 "Some similarities between the linearity of induction of biological effects by radiation and the super-regenerative radio receiver operating in the linear mode." 5 October 1970.

DM-118 "Preliminary basic design for CANDU-OC-1500 Th." N.d.

DM-119 "Thorium fuelled CANDU reactors." N.d.

DM-120 "Energy in the Future – The Role of Nuclear Fission and Fusion." 1971.

DM-121 "Controlled fusion." 5 May 1971.

DM-122 "Thermonuclear fusion and relativistic electron rings." May 1971.

DM-123 "Radioactive Waste Management in the Long Term." 1972.

DM-124 "Brief recollections of Rutherford." June 1971.

DM-125 "Tolerance for carcinogens." July 1971.

DM-126 "Proposed Royal Society of Canada Symposium on Communications into the Home." 20 July 1971.

DM-127 "Global heat balance." 6 August 1971.

DM-128 "Laser plasma interaction for controlled nuclear fusion." 7 October 1971.

DM-129 "Radiation effects in man." N.d.

DM-130 "Royal Society of Canada Symposium on Communication into the Home." 3 November 1971.

DM-131 "Nuclear power development in Canada." 28 November 1971.

DM-132 "Obstetrical X-rays "beneficial" for some?" 18 January 1971.

DM-133 "The Republic of Science." 22 February 1972.

DM-134 "Influence of heterogeneities on alternative interpretations of published studies of pre-natal radiation and childhood cancer, including possible zero carcinogenic effect." 8 November 1972.

DM-135 "Fast breeders are not necessary nor likely ever to meet the economic competition set by CANDU." 12 July 1972.

DM-136 "Basic research needed for CANDU fuel." 14 August 1972.

DM-137 "Canada's opportunity." February 1973.

DM-138 "Why maximum CANDU potential lies with organic coolant and a thorium fuel cycle." 19 January 1973.

DM-139 "On understanding the estimates for large generating stations." N.d.

DM-140 "Nuclear energy based synthetic hydrocarbon fuels – basic technology and economics." 22 May 1973.

Director's Reports

DR-1 "World possibilities for the development and use of atomic power." March 1947.

DR-2 "The future atomic energy pile." September 1947.
DR-3 "The selection, initiation and control of research projects in the Atomic Energy Research Division." 31 January 1948.
DR-4 "Memorandum on progress reports from the Atomic Energy Research Division." 9 March 1948.
DR-5 "Memorandum on selection and promotion boards for NRC staff." 7 April 1948.
DR-6 "Memorandum for the information of those concerned with 'The Irradiation Program.'" 15 June 1948.
DR-7 "Note on the neutron flux in uranium metal in NRX pile." July 1948.
DR-8 "Memorandum re: Export Controls." 29 July 1948.
DR-9 "Unclassified researches in nuclear physics at Chalk River 1948." November 1948.
DR-10 "Plan for the Canadian atomic energy project in the event of war." 10 February 1949.
DR-11 "Memorandum on proposed experiment to detect neutrinos." 16 May 1949.
DR-12 "Notes on the mass of the neutron." December 1949.
DR-13 "Philosophy of design of chemical and chemical engineering laboratories for radioactive work." 2 February 1950.
DR-14 "Interim memorandum on uranium metal for the NRX reactor and future systems." March 1950.
DR-15 "Reflections on reliability and lessons from the Instrumentation Conference at Chalk River." March 1950.
DR-16 "Notes on electrostatic digital storage in cathode-ray tube." 22 November 1951.
DR-17 "Energy stability of ammonium nitrate solutions against explosion." 20 December 1950.
DR-18 "An Atomic Power Proposal." 27 August 1951.
DR-19 "Comparison of PU-240 and U-238 in thermal fission reactors and its relation to the long term irradiation of natural uranium." 10 September 1951.
DR-20 "The Chalk River approach to atomic power." 22 January 1952.
DR-21 "Reactivity Changes Expected and Observed in Long Irradiation of Natural Uranium in the NRX Reactor." 1955.
DR-22 "Chalk River prospects for atomic power (second revision)." 7 January 1957.
DR-23 "Atomic energy objectives for applied nuclear physics research." 2 May 1952.
DR-24 "The significance of the yield of neutrons from heavy nuclei excited to high energy." 25 August 1952.
DR-25 "Memorandum on rod policy for NRU." 11 August 1952.
DR-26 "Activity in effluent water from NRX ruptures." 24 December 1952.

DR-27 "Tentative campaign plan for restoration of NRX." 5 January 1953.
DR-28 "Calculated Reactivity Changes in the Long Irradiation of Natural Uranium." 6 January 1953.
DR-29 "Interim report at January 9, 1953 on problems arising from the NRX incident of December 12, 1952."
DR-30 "Interim note correlating data on the ratio of PU-240 to PU-239 in NRX uranium rods." March 9 1953.
DR-31 "An appreciation of the problem of reactor shut-off rods with special reference to the NRX reactor." May 1953.
DR-32 "The accident to the NRX reactor on December 12, 1952." 13 July 1953.
DR-33 "Note on the possibilities of uranium-plutonium regenerative thermal neutron reactors." 24 August 1953.
DR-34 "Controlling factors in the development of nuclear power." 30 September 1953.
DR-35 "A programme for atomic power in Canada." 15 December 1953.
DR-35b "A programme for atomic power in Canada." 21 March 1955.
DR-36 "Cost of light water as coolant for enriched power reactors." 12 March 1954.
DR-37 "A proposal for a recycled metal power only reactor." 30 August 1954.
DR-38 "Attainable burn-up of uranium with plutonium recycling with special reference to the ANL 1000 MW heavy water boiling reactor of Geneva paper." 3 February 1956.
DR-39 "Low cost fuelling without recycling." 10 December 1956.

AECL REPORTS

203 "Some Economic Aspects of Nuclear Fuel Cycles." *Proceedings of the International Conference on the Peaceful Uses of Atomic Energy* 3, (1956): 3–13.
300 "Atoms for Peace – Canada Goes to Geneva." *Queen's Quarterly* 63 (1956): 86–96.
307 "International Comparisons of Radioactivity Standards." *Nature* 177 (1956): 12–13.
593 "Canada's Steps Towards Nuclear Power." *Proceedings of the Second United Nations International Conference on the Peaceful Uses of Atomic Energy* 1 (1958): 53–9.
651 "Cost Comparisons for Enriched Versus Natural Uranium Fuel and Zirconium Versus Stainless Steel Fuel Sheathing in Bidirectional Slug Fuelled Reactors." 1958.
682 "The Canada-India Reactor: An Exercise in International Cooperation." *Proceedings of the Second United National International Conference on the Peaceful Uses of Atomic Energy* 1 (1958): 355.
704 "Economics of Uranium and Thorium for the Generation of Electricity." September 1958.

797 "Some Highlights of Experience and Engineering of High-Power Heavy-Water-Moderated Nuclear Reactors." October 1959.

799 Paper presented at Atomic Power Symposium, Chalk River, May 1959.

1080 "Competitive Nuclear Power for Canada." *Nucleonics* 18 (1960).

1172 "Research in the Atomic Field." *Canada Year Book* (1960): 415–20.

1318 "Long Term Prospects for Competitive Nuclear Power." Seventh AECL Symposium on Atomic Power, Chalk River, September 1961.

1432 "Research in the Atomic Field." *Canada Year Book* (1961): 378–83.

1580 "Chalk River Picture Book." *International Council of Scientific Unions Review* 4 (1962): 184–97.

1599 "Power Reactor Development Evaluation." Paper presented at the 8th Symposium on Atomic Power, Chalk River, September 1962.

1635 "Fission Fragment Retention." Presidential address, Proceedings of the American Nuclear Society. 1962.

1692 "Heavy Water Power Reactors: Fundamentals of Economic Design." International Conference of the Canadian Nuclear Association, Ottawa, May 1962.

1789 "The Triangle of Economics, Technology and Law." International Conference of the Canadian Nuclear Association, Montreal, May 1963.

1807 "Power Reactor Evaluation and Our Position in the World." Paper presented at the 9th Power Symposium, Toronto, September 1963.

1916 "How Much of the Rocks and Ocean for Power? Exploiting the Uranium-Thorium Fissile Cycle." 1964.

2001 "Electricity Supply in Canada and the Role of Nuclear Power." *Proceedings of the Third United Nations International Conference on the Peaceful Uses of Atomic Energy* 1 (1965): 53–61.

2010 "Prospective D_2O-moderated Power Reactors." *Proceedings of the Third United Nations International Conference on the Peaceful Uses of Atomic Energy* 5 (1965): 333–42.

2019 "Fission-gas Behaviour in UO2 Fuel." *Proceedings of the Third United Nations International Conference on the Peaceful Uses of Atomic Energy* 11 (1965): 405–15.

2100 "The Canada-India Reactor – A Case History." Written with H.J. Bhabha. United Nations Conference on the Application of Science and Technology for the Benefit of the Less Developed Areas, New Delhi, October 1962.

2312 "Experience with Canada's First Nuclear Power Station (NPD) and the Prospects for Heavy Water Power Reactors." *Journal of the British Nuclear Energy Society* 3 (1964): 162–71.

2315 "Canada's Role in Nuclear Power." 1965.

2332 "Canada Holds First to a Single Reactor Trend." *EuroNuclear* 1 (November 1964): 173–74.

2355 "Research in the Atomic Energy Field." *Canada Year Book* (1965): 278–383.

2358 "Research in the Atomic Energy Field." *Canada Year Book* (1963–64): 368–73.
2451 "Discovery of the Electron." *Physics Today* 19, no. 5 (1966): 12–13.
2472 "Research in the Atomic Energy Field." *Canada Year Book* (1966).
2483 "CANDU Reactors to 1980 and in the Long Term." Written with J.L. Gray and L.R. Haywood. N.d.
2486 "Symposium on Atomic Power," no. 11, Toronto. October 1966.
2927 "Nuclear Power – The Next Decade of Development." *Proceedings of the American Power Conference* 29 (1967): 57–62.
2947 "Outlook for Heavy Water Reactors." Paper presented at International Atomic Energy Agency Symposium on Heavy Water Power Reactors, Vienna. September 1957.
3016 "ING: A Vehicle to New Frontiers." *Science Forum* (February 1968): 3–7.
3242 "Power Development in Developing Countries." *Science Reporter* 6 no. 4 (April 1969).
3787 "Designing UO_2 Fuel for Negligible Sheath Strain." 1971.
3980 "Large-Scale Nuclear Energy from the Thorium Cycle." Paper presented at the fourth United Nations International Conference on the Peaceful Uses of Atomic Energy, Geneva. September 1971.
4095 "Performance of Heavy Water and Organic Reactors." *Nuclear News*, 15 October 1971.
4380 "Nuclear Energy and the Quality of Life." *International Atomic Energy Agency Bulletin* 14, no. 4 (1972): 2–14.
4557 "Symposium Summary." The IAEA Symposium on Applications of Nuclear Data in Science and Technology, Paris, March 1973.
4618 "CNA Luncheon Address." Canadian Nuclear Association 13th Annual International Conference, Toronto, June 1973.
4620 "Nuclear Energy the Source for the Future." *Naturwissenschaften* 60 (1973): 501–6.

RETIREMENT

"The Nuclear Energy Industrial Complex and Synthetic Fuels." Sixth Symposium on Energy Resources, The Royal Society of Canada, 189–93. October 1973.
"Energy Cost Prospects." Written for Canadian Press. March 1974.
"Abundant Harnessed Energy at Low Cost and Low Risk from Nuclear Fission." Elizabeth Laird Memorial Lecture, University of Western Ontario, April 1974.
"Nuclear Fission Energy: Essential, Abundant, Lower than ever Cost, and Safe in Capable Hands." *Chemical Technology* (4 September 1971): 531–6.
"Attainable Ratings of CANDU-OC-Thorium Set by the Properties of the Organic Liquid Coolant." *Annals of Nuclear Energy* 2 (1975): 779–86.

"Energy in the Future of Man: From Survival to Super-Living." Lecture, University of Calgary, October 1975.

"The Nuclear Opportunity." *Power Engineering Society Newsletter*, no. s-59 (November 1975).

"The Communication of Science." *Transactions of the Royal Society of Canada* 14 (1976).

"Paradoxes and Popular Misconceptions over Harnessed Nuclear Energy." *Transactions of the Royal Society of Canada* 15 (1977).

"The Impact of Nuclear Energy on Society, Politics and Economics." *Proceedings of the World Conference "Towards a Plan of Action for Mankind"* 2 (1977).

"Facing the Future in an Age of Uncertainty." Convocation address, Laurentian University, Sudbury, Ontario, May 1977.

"New Ideas in Human Evolution." *Transactions of the Royal Society of Canada*, 16 (1978).

"New Prospects for Low-cost Thorium Cycles." *Annals of Nuclear Energy* 5 (1978): 297–304.

"Thorium – The Major Source of Energy." Lecture at the University of Toronto, February 1979.

"Early Detectors and Counters." *Nuclear Instruments and Methods* 162 (1979): 9–14.

Notes

CHAPTER ONE

1 This information comes from a written document supplied by J.A. Lewis during an interview on 22 September 1987 at his home near Heathfield, East Sussex, England.
2 Other works included the Saltash bridge at Plymouth, the Crystal Palace water-towers (demolished after the fire), and a timber slipway for launching the "Great Eastern."
3 Gwynedd Gerry, interview with author, Deep River, Ontario, 8 July 1986.
4 J.A. Lewis, interview with author.
5 This is quoted from a document obtained from J.A. Lewis entitled "Biographical Particulars – Wilfrid Bennett Lewis," n.d.
6 Lewis's early publications include "Note on the Problem of Selectivity Without Reducing the Intensity of the Sidebands," 133–4, and "The Transmitting Station," 261.
7 J.A. Lewis, interview with author.
8 Shoenberg, "W.B. Lewis, 1908–1987"; Lovell and Hurst, "Wilfrid Bennett Lewis," 456–7.
9 Crowther, *The Cavendish Laboratory*, 3.
10 Ibid., 100.
11 Ibid., 102. For more on the Cavendish Laboratory under Thomson see Thomson, *J.J. Thomson and the Cavendish Laboratory*.
12 For a fascinating and detailed examination of Rutherford's life and work see Wilson, *Rutherford*.
13 Ibid., 268.
14 Ibid., 417.
15 Hendry, *Cambridge Physics*, 81.
16 Ibid., 85.
17 Quoted in Lewis, "Frontier Events in Electrical, Electronic and Radio

Engineering From Picowatts to Terawatts in the Last 40 and the next 60 Years," 3 November 1971, CRNL Records, SDDO, AECL 4071.

18 Hendry, *Cambridge Physics*, 50; Hartcup and Allibone, *Cockcroft and the Atom*, 27.

19 Norman Feather, interview, American Institute of Physics Oral History Project, 25 February 1971.

20 Oliphant, *Rutherford Recollections*, 35.

21 Hartcup and Allibone, *Cockcroft*, 27.

22 Hendry, *Cambridge Physics*, 32.

23 Wilson, *Rutherford*, 278–83.

24 Hendry, *Cambridge Physics*, 141.

25 Ibid., 142.

26 Oliphant, *Rutherford Recollections*, 36–7.

27 Ibid., 1.

28 Ibid., 109; Hartcup and Allibone, *Cockcroft*, 31; Crowther, *Cavendish*, 197–8.

29 Kapitza club minutes, Churchill College Archives Centre, Cambridge, Cockcroft Papers, CKFT 7/2–7/3.

30 J.A. Lewis, interview with author; Lovell and Hurst, "Wilfrid Bennett Lewis," 459.

31 A.G. Ward, interview with author, Deep River, Ontario, 3 July 1986.

32 L.G. Cook, interview with author, Summit, New Jersey, 19 November 1987.

33 Hartcup and Allibone, *Cockcroft*, 67.

34 "Biographical Particulars – Wilfrid Bennett Lewis"; Lovell and Hurst, "Wilfrid Bennett Lewis," 461–3.

35 These publications are: Lewis, "The Apparent Demodulation," 538–40; Lewis "The Detector," 487–99; Lewis, "Demodulation," 629–30; Lewis and Milner, "A Portable Duplex Radio-Telephone," 475–82.

36 For a detailed discussion of this device see Lewis, "A Portable Duplex Radio-Telephone," 475–82.

37 Sir J.C. Kendrew, interview with author, Cambridge, England, 5 October 1987; S.W.H.W. Falloon, interview with author, Cambridge, England, 5 October 1987.

38 Lewis, *Electrical Counting*.

CHAPTER TWO

1 Sir Robert Cockburn, interview with author, Fleet, Great Britain, 23 September 1987.

2 For a discussion of the origins of DSIR see MacLeod and Andrews, "The Origins of the D.S.I.R.," 23–48. For an examination of the Board of Invention and Research, the operative arm of wartime research and de-

velopment, see MacLeod and Andrews, "Scientific Advice in the War at Sea," 3–40.

3 For further details on the coordinating boards see Clark, *Tizard*, 57–64.
4 Rowe, *One Story of Radar*, 1–3.
5 Quoted in Jones, *Most Secret War*, 39.
6 On the Air Ministry's ongoing concern with the German air threat throughout the 1930s, see Wark, *The Ultimate Enemy*, 35–79.
7 Rowe, *One Story of Radar*, 1.
8 Clark, *Tizard*, 111.
9 Rowe, *One Story of Radar*, 6.
10 See Watson-Watt, *The Pulse of Radar*, for a biased account of this period. Much more detailed and balanced is Clark, *Tizard*.
11 Quoted in Clark, *Tizard*, 115.
12 Ibid., 115; Rowe, *One Story of Radar*, 9–10.
13 Rowe, *One Story of Radar*, 8.
14 Ibid., 14–15.
15 Postan, Hay, and Scott, *Design and Development*, 377.
16 See Kinsey, *Bawdsey – Birth of the Beam*, for an extensive discussion of the research station at Bawdsey.
17 Postan, Hay, and Scott, *Design and Development*, 376.
18 See Bowen, *Radar Days*, for a detailed account of the development of airborne radar.
19 Rowe, *One Story of Radar*, 24
20 Ibid., 46–8
21 Lovell and Hurst, "Wilfrid Bennett Lewis," 465.
22 Ibid.
23 Hartcup and Allibone, *Cockcroft*, 86.
24 Lovell and Hurst, "Wilfrid Bennett Lewis," 465 fn.
25 Clark, *Tizard*, 171–2; Rowe, *One Story of Radar*, 48; Hartcup and Allibone, *Cockcroft*, 85–6. A further organizational meeting was held in June. See Cockcroft to C.E. Wynn-Williams, 17 June 1939, Churchill College Archives Centre, Cockcroft Papers, CKFT 25/10.
26 Postan, Hay, and Scott, *Design and Development*, 453.
27 See for example Lewis, "Report on an electrically switched receiving aerial having a sharp beam capable of rotation through 360 degrees," 20 September 1939, PRO, AVIA 7/62.
28 J.A. Lewis, interview with author.
29 Ibid.
30 Rowe, *One Story of Radar*, 57; Minutes of meeting of the Scientific Research Board, 2 April 1940 and 3 and 16 April 1941, PRO, AVIA 8/27; Lovell and Hurst, "Wilfrid Bennett Lewis," 479.
31 Donald Watson, interview with R.S. Bothwell, Ottawa, 13 November 1985.

32 Ibid.
33 P.I. Dee to J.D. Cockcroft, 25 May 1940, Churchill College Archives Centre, Cockcroft Papers, CKFT 20/7.
34 Quoted in Lovell and Hurst, "Wilfrid Bennett Lewis," 468; J.C. Kendrew, interview with author.
35 Rowe, *One Story of Radar*, 97.
36 Donald Watson, interview with author, Ottawa, 3 August 1987
37 Sir Robert Cockburn, interview with author.
38 Postan, Hay, and Scott, *Design and Development*, 379–80.
39 Ibid. 378.
40 See Clark, *Tizard*, 248–68 for a complete discussion of the Tizard mission.
41 Lewis, memo entitled "Notes on Reports from British Technical Mission to U.S.A.," 17 November 1940, PRO, AVIA 7/78.
42 Lovell and Hurst, "Wilfrid Bennett Lewis," 478.
43 Lewis to Chairman, British Radar Mission, 23 October 1943, PRO, AVIA 7/2253.
44 Lewis to D.H. Johnson, 15 February 1944, PRO, AVIA 7/2253.
45 Postan, Hay and Scott, *Design and Development*, 453.
46 Ibid., 440.
47 Ibid., 479.
48 Ibid., 480–1.
49 E.H. Putley, interview with author, Malvern, England, 14 October 1987.
50 Lovell and Hurst, "Wilfrid Bennett Lewis," 475.
51 Memo by Lewis, 14 May 1945, PRO, AVIA 7/846.
52 Ibid.
53 Lovell and Hurst, "Wilfrid Bennett Lewis," 479–80.
54 Lewis, "The Role of TRE in the National Scientific Effort," 1, PRO, AVIA 15/2260.
55 Quoted in Lovell and Hurst, "Wilfrid Bennett Lewis," 482.
56 Quoted in ibid., 483.
57 Lewis, "The M. of S.(Air) Radio Establishment and Its Relations with Headquarters, the Users, Industry, Universities and Other Research Organisations," 3 May 1946, RSRE, L/M55/WBL.
58 Lovell and Hurst, "Wilfrid Bennett Lewis," 481.
59 "Biographical Particulars – Wilfrid Bennett Lewis."
60 Lovell and Hurst, "Wilfrid Bennett Lewis," 484.

CHAPTER THREE

1 Donald Watson, inteview with author, 3 August 1987.
2 Wilson, *Rutherford*, 5.
3 See Eggleston, *National Research in Canada*, for further details.
4 Senate, *A Science Policy for Canada*, vol. 1, 61.

5 An explanation of the term "isotope" is required here. A nucleus is made up of protons and neutrons, e.g., U-238 has 92 protons and 146 neutrons (92 + 146 = 238). Surrounding the nucleus are electrons, equal in number to the protons, making the atom as a whole electrically neutral. The number of electrons determines the chemical properties, so that all atoms of the same chemical element must have the same number of protons in their nuclei. But they can have different numbers of neutrons, and therefore different weights. Atoms with the same number of protons but differing numbers of neutrons are called "isotopes."
6 Weart, *Scientists in Power*, 127–34.
7 Ibid., 153–67.
8 Gowing, *Britain and Atomic Energy*, 39.
9 Ibid., 51.
10 Ibid., 65–7.
11 Ibid., 123–32.
12 Ibid., 71; Bothwell, *Nucleus*, 12–14.
13 Bothwell, *Nucleus*, 21.
14 For details on this breakdown in cooperation see ibid., 33–6, and Gowing, *Britain and Atomic Energy*, 147–77.
15 Bothwell, *Nucleus*, 79.
16 Ibid., 80.
17 Telegram no. 344 from High Commissioner in Canada to Secretary of State for Dominion Affairs, 25 February 1946, PRO, FO 800/582.
18 Telegram no. 548 from PM to Lord Halifax and Field Marshal Wilson, 6 February 1946, PRO, FO 800/583.
19 Gowing, *Independence and Deterrence*, vol. 1, 136; George Laurence, interview with R.S. Bothwell, Deep River, Ontario, 24 July 1985.
20 Gowing, *Independence and Deterrence*, vol. 2, 206.
21 Eggleston, *Canada's Nuclear Story*, 121; Bothwell, *Nucleus*, 116–17.
22 C.G. Stewart, interview with author, Deep River, Ontario, 11 July 1986.
23 Chalk River Research Staff to Minister through the president of the NRC, 30 September 1947, CRNL Records, 1500/AECL-overall, vol. 1.
24 Mackenzie to Keys, 10 October 1947, NA, C.J. Mackenzie Papers, MG 30 B122, vol. 5.
25 Gwen Milton, interview with author, Chalk River, Ontario, 12 April 1989.
26 Mackenzie to Cockcroft, 22 January 1947, NA, Mackenzie Papers, MG 30 B122, vol. 5, Atomic Energy, vol. 2.
27 Gowing, *Independence and Deterrence*, vol. 2, 207.
28 National Research Council, *Review, 1949*, 54.
29 Eggleston, *Canada's Nuclear Story*, 288.
30 Ibid., 286–7.
31 Ibid., 289–91.

32 The details of research performed by the chemistry branch can be found for the years 1947–52 in National Research Council, *Review*.
33 Information for this section was found in National Research Council, *Review* for 1947–52, and Marko, Butler, and Myers, "Biochemistry at Chalk River," 14–19.
34 A.M. Marko, interview with author, Chalk River, Ontario, 29 July 1987.
35 Lewis, "The importance of fundamental research in AECL," 22 January 1969, CRNL Records, SDDO, DM-101.
36 Lewis, "Atomic energy objectives for applied nuclear physics research," 2 May 1952, CRNL Records, SDDO, DR-23; Lewis, "The significance of the yield of neutrons from heavy nuclei excited to high energy," 25 August 1952, DR-24.
37 See Lewis, "Atomic energy objectives for applied nuclear physics research," 2 May 1952, CRNL Records, SDDO, DR-23, and Lewis, "Controlled H-fusion," 29 January 1958, DM-50.
38 Eggleston, *Canada's Nuclear Story*, 294.
39 Zimmerman, *The Great Naval Battle of Ottawa*, 26; Bothwell, *Nucleus*, 19–22 and 63–5. For more complimentary descriptions of Mackenzie's administrative abilities see Eggleson, *National Research in Canada*, and King, *E.W.R. Steacie and Science in Canada*.
40 Donald Watson, interview with author, 3 August 1987.
41 Report by J.D. Cockcroft, "Cooperation in Atomic Energy between Canada and the United Kingdom," 10 September 1946, CRNL Records, 1615 F1, vol. 1.
42 Gowing, *Independence and Deterrence*, vol. 1, 104–6.
43 Mackenzie to Lewis, 14 February 1947, NA, Mackenzie Papers, MG 30 B122, vol. 5, Atomic Energy, vol. 2.
44 Gowing, *Independence and Deterrence*, vol. 2, 10.
45 Cockcroft to Mackenzie, 25 November 1946, PRO, FO 800/582.
46 Lewis to Mackenzie, 13 November 1946, CRNL Records, 1800 UK/overall, vol. 1; Mackenzie to Lewis, 21 December 1946, ibid.
47 Thewlis to Lewis, 18 July 1947; Lewis to Cockcroft, 1 August 1947; Bretscher to Lewis, 15 August 1947; Lewis to Bretscher, 27 August 1947, ibid.
48 H.W.B. Skinner to Lewis, 10 January 1947, ibid. There was a series of correspondence through this year between Lewis, Cockcroft, and various branch heads about this problem.
49 Cockcroft to Lewis, 6 November 1946, PRO, AB6/140.
50 Gowing, *Independence and Deterrence*, vol. 1, 323.
51 Hewlett and Duncan, *Atomic Shield*, 280.
52 Colonel Curtis Nelson, interview with R.S. Bothwell, Washington, 17 April 1985.
53 C.J. Mackenzie to Carroll Wilson, 14 October 1949, NA, Mackenzie Papers, MG 30 B122, vol. 5.

54 In fact, NRX was not finally shut down until April 1993, nearly forty-six years after it began operating.
55 Minutes of Future Systems Committee meeting, 7 November 1947, CRNL Records, SDDO.
56 Ibid.
57 Ibid.
58 Minutes of Future Systems Committee meeting, 26 January 1948, CRNL Records, SDDO.
59 Minutes of Future Systems Committee meeting, 15 April 1948, CRNL Records, SDDO.
60 Quoted in Bothwell, *Nucleus*, 76.
61 Minutes of special meeting of Future Systems Committee with Atomic Energy Control Board, 29 May 1948, CRNL Records, SDDO.
62 K.F. Tupper to Vice-president and Director, Division of Research, 18 November 1948, CRNL Records, 6000 NRU-overall, vol. 1.
63 Memorandum by Lewis, "Policy for Future Pile at Chalk River," 3 February 1949, CRNL Records, 6000 NRU-overall, vol. 1.
64 Minutes of Future Systems Committee meeting, 11 February 1949, CRNL Records, SDDO.
65 Minutes of Future Systems Committee meeting, 29 March 1949, CRNL Records, SDDO.
66 Reactor energy outputs are described either as heat (megawatts thermal, MWt) or as electricity (megawatts electric, MWe). Generally speaking, for a given power reactor, the output in MWe is between one-quarter and one-third of the output in MWt.
67 Minutes of Future Systems Committee meeting, 28 June 1949, CRNL Records, SDDO.
68 Minutes of Future Systems Committee meeting, 21 July 1949, CRNL Records, SDDO.
69 Minutes of Future Systems Committee meeting, n.d., CRNL Records, SDDO.
70 Laurence to Lewis, 11 January 1950, CRNL Records, 6000 NRU-overall, vol. 1.
71 Mackenzie to Wilson, 20 April 1950, NA, Mackenzie Papers, MG 30 B122, vol. 5.
72 Minutes of meeting to discuss the program for the NRU pile, 29 September 1950, CRNL Records, 1615 N13, vol. 1.
73 Lewis to Gray, 12 August 1952, CRNL Records, 6000 NRU-overall, vol. 5; Laurence to Gray and Lewis, 9 December 1952; B.A. Culpeper to Gray, 13 December 1952; Lewis to Gray, 17 December 1952; MacKay to Laurence, 9 January 1953; Laurence to Gray, 13 January 1953, ibid., vol. 6.
74 Report by J.L. Gray on NRU – July 1950 to January 1954, 20 January 1954, CRNL Records, 6000 NRU-overall, vol. 11.

CHAPTER FOUR

1 National Research Council, *Review, 1948*, 46.
2 Alec Eastwood, interview with author, Deep River, Ontario, 12 August 1987.
3 A.G. Ward, interview with author.
4 Hank Clayton, interview with author, Deep River, Ontario, 18 July 1986; John M. Pawson to Lewis, 10 March 1965, Queen's University Archives, Wilfrid Bennett Lewis Papers, Box 2, file correspondence 1965.
5 Minutes of Board of Management of the Deep River Library, 21 May 1947, W.B. Lewis Public Library Records.
6 Hank Clayton, interview with author.
7 Ibid.
8 Annual General Meeting, 21 January 1958, W.B. Lewis Public Library Records.
9 Minutes of Special General Meeting, 19 October 1966, W.B. Lewis Public Library Records.
10 Minutes of Board of Management Meeting, 15 April 1954, W.B. Lewis Public Library Records.
11 Lewis, "On Public Library Service and its Aims," 16 August 1972, W.B. Lewis Public Library Records.
12 E.C.W. Perryman, interview with author, Deep River, Ontario, 25 June 1986.
13 Gwynedd Gerry, interview with author.
14 E.C.W. Perryman, interview with author.
15 Harry Collins, interview with R.S. Bothwell, Point Alexander, 7 July 1986; George Pon, interview with R.S. Bothwell, Ottawa, 20 October 1986.
16 Donald Hurst, interview with author, Chalk River, Ontario, 20 August 1987.
17 Boyer, *By the Bomb's Early Light*, 113–17.
18 Lewis, "The Role of TRE in the National Scientific Effort," 9 October 1945, PRO, AVIA 15/2260.
19 Lewis, "Possible Future Projects," 14 October 1946, CRNL Records, WBL Papers – miscellaneous.
20 Gowing, *Independence and Deterrence*, vol. 2, 263.
21 C.H. Secord, "Heat and Power from Nuclear Energy," September 1946, CRNL Records, WBL Papers – miscellaneous.
22 Memo from Lewis to H. Clayton, L.G. Cook, D.G. Hurst, K. Tupper, B.W. Sargent and D. Watson, 29 November 1946, CRNL Records, WBL Papers – miscellaneous.
23 Minutes of meeting to discuss Secord report, 9 December 1946, CRNL Records, WBL Papers – miscellaneous.
24 Lewis to Cockcroft, 10 December 1946, CRNL Records, WBL Papers – miscellaneous.

25 Ibid.
26 Lewis, "World possibilities for the development and use of atomic power," 25 March 1947, 2, CRNL Record, SDDO, DR-1.
27 Ibid.
28 Ibid., 4.
29 Ibid., 3.
30 Ibid., 7–9.
31 Lewis, "The future atomic energy pile," 8 September 1947, 2, CRNL Records, SDDO, DR-2. Lewis's mention of using oxide instead of metal is interesting in view of the fact that he was one of the last to switch over to the idea of fuelling NPD with uranium dioxide instead of uranium metal in 1955. See Bothwell, *Nucleus*, 230.
32 Lewis to Mackenzie, 11 September 1947, CRNL Records, 6000 NRU-overall, vol. 1.
33 Laurence to Lewis, 11 September 1947, CRNL Records, 6000 NRU-overall, vol. 1.
34 Memo from Laurence to Future Systems Group, 2 April 1948, CRNL Records, 1615 F1 Future Systems Group, vol. 1; ibid., memo from Hurst to Future Systems Group, 5 April 1948.
35 Mackenzie to Austin Wright, 3 March 1948, NA, Mackenzie Papers, MG 30 B122, vol. 5, Atomic Energy, vol. 3.
36 Lewis, "The Gleam in the Eye of the Atomic Scientist," 15 November 1949, CRNL Records, SDDO, DL-6.
37 Laurence to Lewis, "The Importance of Proposed New Reactor as a Step towards Atomic Power in Canada," 30 March 1950, CRNL Records, 6000 NRU-overall, vol. 1.
38 Ibid.
39 Lewis, "An Atomic Power Proposal," 27 August 1951, 1–3, CRNL Records, SDDO, DR-18.
40 Ibid., 4–8.
41 Lewis, "The Chalk River approach to atomic power," 21 January 1952, CRNL Records, SDDO, DR-20.
42 Ibid.

CHAPTER FIVE

1 Bothwell, *Nucleus*, 143–5.
2 Lewis, "The accident to the NRX reactor on December 12, 1952," 13 July 1953, CRNL Records, SDDO, DR-32.
3 Bothwell, *Nucleus*, 163.
4 P.R. Tunnicliffe, interview with author, Deep River, Ontario, 14 July 1986.
5 Laurence, "Nuclear Reactor Safety in Canada," 20.
6 Hinton to Lewis, 28 February 1952, CRNL Records, 1800 UK/overall, vol. 1.
7 Ibid.

8 Minutes of meeting of Nuclear Physics and Reactor Engineering Group, 2 July 1952, CRNL Records, 1615 F1 Future Systems Group, vol. 1.
9 Minutes of meeting of Metallurgy and Chemical Processing Group, 25 July 1952, CRNL Records, 1615 F1 Future Systems Group, vol. 1.
10 Minutes of meeting of Nuclear Physics and Reactor Engineering group, 19 August 1952, CRNL Records, 1615 F1 Systems Group, vol. 1.
11 MacKay to Lewis and Laurence, 19 August 1952, CRNL Records, WBL Papers – miscellaneous.
12 Lewis, "Summarized Recommendations for Presentation to Board of Directors Meeting 30 September 1952", 29 September 1952, CRNL Records, 1615 F1 Future Systems Group, vol. 1. Note that the proposed applied research into the electrical generation of fissile material is an early mention of an idea that would reappear in the 1960s as the Intense Neutron Generator.
13 Board of Directors minutes, 29–30 September 1952.
14 Lewis, "Chalk River Research Review and Prospect February 1953," 18 February 1953, CRNL Records, SDDO, DL-10.
15 Lewis, "The Goose that Lays the Golden Eggs", 13 March 1953, CRNL Records, SDDO, DL-11.
16 Lewis to Jarvis, 12 August 1953, AECL Records, 103-A-2, Advisory Committee on Atomic Power, vol. 1.
17 Bothwell, *Nucleus*, 191–3; John Foster, interview with author, Mississauga, 25 April 1988.
18 Lewis, "Feasibility Study for Pilot Atomic Power Reactor", 3 November 1953, CRNL Records, WBL Papers – miscellaneous; Minutes of meeting to discuss report, 6 November 1953, ibid.
19 Minutes of meeting of Power Reactor Group, 3 March 1954, CRNL Records, 1615 P14 Power Reactor Group, vol. 1.
20 Minutes of second meeting of Power Reactor Group, 6 April 1954, CRNL Records, 1615 P14 Power Reactor Group, vol. 1.
21 Lewis, "Small power reactor design for pressurized heavy water natural uranium 50 megawatt reactor," 6 April 1954, CRNL Records, SDDO, DM-23.
22 Minutes of fifth meeting of Power Reactor Group, 14 July 1954, CRNL Records, 1615 P14 Power Reactor Group, vol. 1.
23 Minutes of meeting of Power Reactor Group, 3 March 1954, CRNL Records, 1615 P14 Power Reactor Group, vol. 1.
24 Minutes of third meeting of Power Reactor Group, 25 May 1954, CRNL Records, 1615 P14 Power Reactor Group, vol. 1.
25 Minutes of fifth meeting of Power Reactor Group, 14 July 1954, CRNL Records, 1615 P14 Power Reactor Group, vol. 1.
26 Ibid.
27 Ibid.

28 Meeting to discuss proposed program for introducing full-scale economic generation of nuclear power, 26 July 1954, CRNL Records, WBL Papers – miscellaneous.
29 Bothwell, *Nucleus*, 203–4.
30 Ibid.
31 Minutes of meeting of Power Reactor Group, 17 August 1954, CRNL Records, 1615 P14 Power Reactor Group, vol. 1.
32 Minutes of meeting of Power Program Policy Committee, 3 December 1954, AECL Records, C-103-P-1, Power Program Policy Committee, vol. 1.
33 Ibid.
34 J.A.L. Robertson, interview with author, Deep River, Ontario, 21 July 1987; Eugene Critoph, interview with author, 25 July 1987.
35 CRNL Records, 1615 P14 Power Reactor Group vol. 1, minutes of meeting of Power Reactor Group, 31 January 1955.
36 Lewis to Bennett, 17 March 1955; Bennett to Howe, 23 March 1955, AECL Records, 105-6-1 Reactors-NDP, vol. 1.
37 Minutes of meeting at CGE, 13 May 1955, CRNL Records, 6000 NPD-overall, vol. 1.
38 Smith to Lewis, 4 July 1955, AECL Records, 105-6-1 Reactor NPD, vol. 1; Gray to Lewis, 4 July 1955, ibid.; Minutes of meeting of NPD Project committee, 6 July 1955, CRNL Records, 6000 NPD-overall, vol. 1; Bennett to Howe, 28 July 1955, NA, C.D. Howe Papers, MG 27 III B20, vol. 10 S-8-2 folder 18.
39 Lewis, "A programme for atomic power in Canada", 21 March 1955, CRNL Records, SDDO, DR-35b.
40 Report entitled "Atomic Power in Canada", Economics branch, Department of Trade and Commerce, December 1954, NA, Howe Papers, MG 27 III B20, vol. 10 S-8-2 folder 18.
41 Report entitled "The Economics of Atomic Power Reactors", Economics branch, Department of Trade and Commerce, January 1955, NA, Howe Papers, MG 27 III B20, vol. 10 S-8-2 folder 18.
42 For examples of Lewis's economic arguments in favour of atomic power see Lewis, DL-16, 8 March 1955; DL-17, 12 April 1955; DL-18, May 1955.
43 Minutes of NPD Project Committee, 14 October 1955, CRNL Records, 6000 NPD-overall, vol. 1.
44 John Foster, "The First Twenty-five Years," speech to the Canadian Nuclear Association, June 1985, 14.
45 Minutes of meeting of Power Program Policy Committee, 29 February 1956, AECL Records, C-103-P-1, Power Program Policy Committee.
46 Ibid.
47 Minutes of meeting of the NPD Coordinating committee, 6 December 1956, CRNL Records, 6000 NPD-overall, vol. 2.

48 Siddall to Members of the NPD Technical committee, 11 December 1956, CRNL Records, 6000 NPD-overall, vol. 2.
49 Tunnicliffe to NPD Technical committee, 20 December 1956, CRNL Records, 6000 NPD-overall, vol. 2.
50 Foster to Lewis, 28 December 1956, CRNL Records, 6000 NPD-overall,vol. 2.
51 Minutes of meeting of the NPD Coordinating committee, 1 February 1957, CRNL Records, 6000 NPD-overall, vol. 3.
52 Smith to Bennett, 25 March 1957, AECL Records, 105-6-1 Reactors NPD, vol. 1.
53 Harold Smith, interview with author, Mississauga, 2 May 1988.
54 Lewis, "The Case for NPD," 26 March 1957, 2, CRNL Records, SDDO, DM-43 (draft).
55 Executive Committee minutes, 27 March 1957.
56 John Foster, interview with author, 25 April 1988.
57 Lewis to Bennett, 23 December 1957, AECL Records, 105-6-1 Reactors NPD, vol. 2.
58 Minutes of meeting of NPD Coordinating Committee, 6 February 1958, CRNL Records, 6000 NPD-overall, vol. 5.

CHAPTER SIX

1 R.M. Fishenden to Lewis, 8 September 1952; Lewis to Fishenden, 23 September 1952, CRNL Records, 1802-UK/1 vol. 1.
2 Lewis to Cockcroft, 17 December 1952, CRNL Records, 1802-UK/1 vol. 1.
3 Report by H. Sheard on discussions on Chalk River irradiations held with W.B. Lewis and J.D. Cockcroft, 4 October 1954, PRO, AB6 1478.
4 Telegram Hinton to Plowden, 20 September 1954, PRO, AB19/13.
5 Meeting between Canada and United Kingdom, 28 September 1954, PRO, AB16/255.
6 Hinton to Lewis, 19 May 1955, PRO, AB 19/14.
7 Lewis to Laurence, 24 September 1955, CRNL Records, 5001 CANDU Reactor Concepts, vol. 1; Lewis to Bennett, 26 September 1955; Bennett to Lewis, 25 October 1955, AECL Records, 105-3 Large Reactor Program (CANDU), vol. 1.
8 Hinton to Cockcroft, 28 September 1955, PRO, AB 19/16.
9 For an interesting examination of the development of power reactors in the United States see Dawson, *Nuclear Power*.
10 Dawson, *Nuclear Power*, 37.
11 John Melvin, interview with R.S. Bothwell, Chalk River, 26 September 1985.
12 Text of "Agreement for Cooperation concerning Civil Uses of Atomic Energy between the Government of the United States of America and the Government of Canada," 2 May 1955, NA, Howe Papers, MG 27 III B20, vol. 10 S-8-2 folder 18.

13 Eayrs, *Canada in World Affairs*, vol. 7, 205.
14 English, *The Wordly Years*, 35–8.
15 For a highly biased look at the life of Homi Bhabha see Mishra, *Five Eminent Scientists*, and Kulkarni, *Homi Bhabha*; a balanced view is in Bothwell, *Nucleus*, 351–2. For a critical discussion of Bhabha's influence on Indian science policy see Sharma, "India's Lopsided Science," 32–5.
16 W.J. Bennett speech, University of Toronto, 11 January 1989; Lewis to Bhabha, 5 April 1955, CRNL Records, 6000 CIR-overall, vol. 1.
17 Gwynedd Gerry, interview with author.
18 Lewis to Bhabha, 29 april 1955, CRNL Records, 6000/CIR-overall, vol. 1.
19 Mishra, *Five Eminent Scientists*, 162.
20 Bennett to Ritchie, 21 March 1955, DEA Records, 11038-1-40.
21 Quoted in Watson to C.C. Kennedy, 9 April 1955, CRNL Records, 6000 CIR-overall, vol. 1.
22 Lewis, "International Aspects of Atomic Energy," 12 September 1955, CRNL Records, SDDO, DM-33.
23 Lewis and Bhabha, "The Canada-India Reactor: An Exercise in International Cooperation," December 1958, 4, CRNL Records, SDDO, AECL-682.
24 Ibid., 3.
25 In Bothwell, *Nucleus*, 357.
26 Lovell and Hurst, "Wilfrid Bennett Lewis," 498.
27 Pilat, Pendley, and Ebinger, eds., *Atoms for Peace*, 6.
28 Resolution adopted by the General Assembly of the United Nations at its 503d plenary meeting on 4 December 1954, CRNL Records, 1773-2 vol. 1.
29 Goldschmidt, *The Atomic Complex*, 257–8; Lewis, "Atoms for Peace," 86–96.
30 Lewis, "Atoms for Peace," 89.
31 Minutes of meeting held 28 September 1956, CRNL Records, 1773-2 vol. 1, Cooperation and Liaison UNSAC.
32 D.V. LePan to Lewis, 10 November 1958, CRNL Records, 1773-2 vol. 3, Cooperation and Liaison UNSAC.
33 Telegram Vienna to External Affairs, 30 September 1958; Lewis to Sterling Cole, 1 October 1958, CRNL Records, 1771-2 vol. 1.
34 Lewis to N.A. Robertson, 24 March 1960, CRNL Records, 1773-2 vol. 3, Cooperation and Liaison UNSAC.
35 E.W.R. Steacie to N.A. Robertson, 28 January 1960, CRNL Records, 1773-2 vol. 3, Cooperation and Liaison UNSAC.
36 Lewis to Cockcroft, 4 March 1960, CRNL Records, 1773-2 vol. 3, Cooperation and Liaison UNSAC.
37 Lewis to N.A. Robertson, 24 March 1960, CRNL Records, 1773-2 vol. 3, Cooperation and Liaison UNSAC.

38 Summary notes of meeting of group of experts on the effects and implications of nuclear weapons, 6 March 1967, CRNL Records, 1773-5 vol. 4.
39 Notes on meeting of group of consultant experts on the effects and implications of nuclear weapons, 5 July 1967, CRNL Records, 1773-5 vol. 4.
40 Quoted in Lovell and Hurst, "Wilfrid Bennett Lewis," 497. Lewis donated his part of the honorarium to McGill University to support the development of a superconducting magnet.
41 D. Watson, interview with author, 3 August 1987; G.C. Laurence, interview with author, 12 July 1986.

CHAPTER SEVEN

1 Minutes of Future Systems Committee, 1 October 1963, CRNL Records, 1615/F1 vol. 2.
2 Minutes of meeting of Future Systems committee, 18 December 1963, CRNL Records, 1615/F1 vol. 3.
3 Elliott, Lewis, and Ward, "ING: A Vehicle to New Frontiers," 4.
4 Minutes of first meeting of High Neutron Flux Facility Study Committee, 30 July 1963; minutes of third meeting of High Neutron Flux Facility Study Committee, 28 August 1963, CRNL Records, 9226-2-3-1, vol. 1, ING Study Committee.
5 Minutes of second meeting of High Neutron Flux Facility Study Committee, 14 August 1963; minutes of sixth meeting of High Neutron Flux Facility Study Committee, 6 November 1963, CRNL Records, 9226-2-3-1, vol. 1, ING Study Committee.
6 Minutes of fourth meeting of High Neutron Flux Facility Study Committee, 19 September 1963, CRNL Records, 9226-2-3-1, vol. 1, ING Study Committee.
7 Minutes of meeting of Senior Management Committee, 15 May 1964, AECL Records, 103-S-8 vol. 1.
8 Lewis to J.L. Gray and board of directors, 28 May 1964, CRNL Records, 9226-1-4, vol. 2, ING general.
9 Ibid.
10 Minutes of eleventh meeting of High Neutron Flux Facility Study Committee, 9 June 1964, CRNL Records, 9008-3 vol. 1.
11 Minutes of eleventh meeting of High Neutron Flux Facility Study Committee, 9 June 1964, CRNL Records, 9008-3 vol. 1.
12 Minutes of third meeting of ING Policy Committee, 16 February 1965, CRNL Records, 9226-2-2-1 vol. 1.
13 Minutes of ninth meeting of ING Policy Committee, 17 December 1965, CRNL Records, 9226-2-2-1 vol. 1.
14 Minutes of eleventh of ING Policy Committee, 14 April 1966, CRNL Records, 9226-2-2-1 vol. 1.

15 Minutes of thirteenth meeting of ING Policy Committee, 28 July 1966, CRNL Records, 9226-2-2-1 vol. 1.
16 Minutes of second meeting of ING Policy Committee, 24 November 1964, CRNL Records, 9226-2-2-1 vol. 1.
17 Minutes of seventh meeting of ING Policy Committee, 6 August 1965, CRNL Records, 9226-2-2-1 vol. 1.
18 Lewis, "Presentation of ING proposal to the Privy Council Committee and Other Government Representatives," 25 August 1966, CRNL Records, 9226-2-1-2 vol. 2.
19 Minutes of third meeting of ING Policy Committee, 16 February 1965, CRNL Records, 9226-2-2-1 vol. 1.
20 Memorandum from W.B. Lewis to Board of Directors, "The aims and desirability of the continued growth of AECL," August 1964, CRNL Records, SDDO, DM-77.
21 D.D. Stewart to J.W. Tomecko, 25 February 1965, CRNL Records, 9008-2 vol. 1.
22 W.R. Livingston to L.R. Haywood, 23 September 1965, CRNL Records, 9008-1 vol. 1.
23 Lewis, "Presentation of ING Proposal to the Privy Council Committee and Other Government Representatives," 25 August 1966, CRNL Records, 9226-2-1-2 vol. 2. Lewis's emphasis.
24 Minutes of the seventh ING Policy Committee, 6 August 1965, CRNL Records, 9226-2-2-1 vol. 1.
25 Minutes of ING Study Advisory Committee, 9 November 1965, CRNL Records, 9226-2-1-1 vol. 1.
26 Memo to Lewis and members of the ING Study Advisory Committee from Larkin Kerwin, 23 December 1965, CRNL Records, 9226-2-1-2 vol. 1.
27 Gray to presidents of universities, 14 September 1965, CRNL Records, 9226-2-2-2 vol. 1.
28 L.E.H. Trainor to W.B. Lewis, 6 July 1965; Lewis to Trainor, 15 July 1965, CRNL Records, 7301-1-Toronto vol. 1.
29 Minutes of first meeting of ING Study Advisory Committee, 9 November 1965, CRNL Records, 9226-2-1-1 vol. 1.
30 Ibid.
31 ING Study Advisory Committee, general correspondence, 23 December 1965, CRNL Records, 9226-2-1-2 vol. 1.
32 Memo from L. Katz to members of ING Study Advisory Committee, 22 January 1966, CRNL Records, 9226-2-1-2 vol. 1.
33 Minutes of fourth meeting of ING Policy Committee, 14 April 1965, CRNL Records, 9008-3 vol. 1.
34 R.E. Grant to R.D. Sage, 2 December 1965, CRNL Records, 9008-2 vol. 2.
35 Memo from Shawinigan Engineering Company on ING design-study reference-site selection, 15 December 1965, CRNL Records, 9226–2–2–2 vol. 1.

36 Lewis, draft of paper "Location of ING", n.d., CRNL Records, 9226-2-2-2 vol. 1.
37 Minutes of second meeting of Study Advisory Committee, 17 January 1966, CRNL Records, 9226-2-1-1 vol. 1.
38 Minutes of fourth meeting of Study Advisory Committee, 23 April 1966, CRNL Records, 9226-2-1-1 vol. 1.
39 R.D. Sage, "ING Reference Site at CRNL," 5 July 1966, CRNL Records, 9008–1 vol. 2.
40 Roland Prevost, "Scientific Decolonisation," *La Presse*, 26 and 28 August 1967, CRNL Records, 9008-2 vol. 7.
41 Trip report of J.A. Quarrington, 11–15 December 1967, CRNL Records, 7301-1-1.
42 Gray to Lewis, 23 February 1967, CRNL Records, 9008-1 vol. 3.
43 Minutes of fifteenth meeting of ING Policy Committee, 11 November 1966, CRNL Records, 9226-2-2-1 vol. 1.
44 Quoted in Canada, Senate, Special Committee, *A Science Policy for Canada*, 101.
45 Canada, Senate, Special Committee, *Proceedings*, 30 and 31 October 1968.
46 Canada, Privy Council Office, Science Secretariat, "The Proposal for an Intense Neutron Generator," xi.
47 Ibid., xii.
48 "Needs before status," *Globe and Mail*, 15 July 1967.
49 Elliott, Lewis, and Ward, "ING: A Vehicle to New Frontiers," 3–7.
50 Parr, "ING: The wrong thing, in the wrong place, at the wrong time," 8–9.
51 McNeill, "ING: A good thing – under certain conditions," 11.
52 David Spurgeon, "The generator that is producing intense reactions," *Globe and Mail*, 20 July 1967.
53 Letter to the Editor, *Globe and Mail*, 28 August 1968.
54 Memo Lewis to all members and alternates of AECL, 4 October 1968, CRNL Records, 9008-1 vol. 20.
55 Haywood to Gray, 25 October 1968, CRNL Records, 9008-1 vol. 20.
56 Minutes of meeting of Senior Management Committee, 5 September 1968; minutes of meeting of Senior Management Committee, 3 October 1968, CRNL Records, S2/overall vol. 7.
57 For a detailed look at this debate over the Queen Elizabeth II telescope see Jarrell, *Cold Light of Dawn*.
58 This disillusionment with science in America is discussed in detail in Kevles, *The Physicists*, 410–26.
59 Doern, *Science and Politics in Canada*, 102.
60 Canada, Senate, Special Committee *Proceedings*, 691.
61 Lewis to J.H. Chapman, 22 February 1973, Queen's University Archives, Wilfrid Bennett Lewis Papers, Box 1.

CHAPTER EIGHT

1 For a detailed examination of the development of power reactors in Canada in this period see Bothwell, *Nucleus*, 279–345.
2 See Bothwell, *Nucleus*, 300–45.
3 This research is discussed in greater detail in chapter 3.
4 Bothwell, *Nucleus*, 262; M.J. McNelly, "A Heavy-Water Moderated Power Reactor."
5 AECL Board of Directors minutes, 10 November 1958.
6 Bothwell, Drummond, and English, *Canada since 1945*, 200.
7 AECL Executive Committee minutes, 20 February 1959.
8 AECL Executive Committee meeting, 25 June 1959.
9 Gray to Gordon Churchill, 14 July 1959, AECL Records, 101–13.
10 Bothwell, *Nucleus*, 269–70.
11 Minutes of Power Reactor Development Programme Evaluation Committee (PRDPEC), appendix A, 11/6/65, AECL Records, 103-P-1.
12 Minutes of meeting of PRDPEC, 23 March 1962; summary report of Power Reactor Development Evaluation Committee, 11 April 1963, AECL Records, C-103-P-1.
13 Lewis, "Report on Power Reactor Development Evaluation," 14 March 1963, AECL Records, C-103-P-1.
14 Ibid.
15 Ibid.
16 Minutes of meeting of PRDPEC, 21 March 1963, AECL Records, C-103-P-1.
17 Draft report "Power Reactor Development Evaluation," 11 April 1963, AECL Records, C-103-P-1.
18 Minutes of meeting of PRDPEC, 7 Januray 1964, AECL Records, C-103-P-1.
19 Haywood to Gray, Foster, and Lewis, 27 January 1964, AECL Records, 105–22, Pickering, vol. 1.
20 Memo by Lewis, 29 January 1964, AECL Records, 105–22, Pickering, vol. 1.
21 Haywood to Gray, 3 February 1964, AECL Records, 105–22, Pickering, vol. 1.
22 Minutes of PRDPEC meeting, 7 January 1964, AECL Records, C 103-P-1.
23 Minutes of PRDPEC meeting, 24 March 1964, AECL Records, C 103-P-1.
24 Foster to Haywood, 23 March 1966, AECL Records, C 103-P-1.
25 Minutes of PRDPEC meeting, 28 April 1966, AECL Records, C 103-P-1.
26 See for example Lewis, "Economics of Uranium and Thorium for the Generation of Electricity," 7–11 September 1958, CRNL Records,

SDDO, AECL 704; Lewis, "How Much of the Rocks and the Oceans for Power? Exploiting the Uranium-Thorium Fission Cycle," April 1964, AECL 1916.
27 Lewis, "Economics of Uranium and Thorium for the Generation of Electricity," 7–11 September 1958, CRNL Records, SDDO.
28 Quoted in Lovell and Hurst, "Wilfrid Bennett Lewis," 492.
29 Ibid.
30 Minutes of meeting of PRDPEC, 2 October 1968, AECL Records, 103-P-1.
31 AECL Board of Directors minutes, 14 August 1969.
32 AECL Executive Committee minutes, 18 January 1973.
33 Ibid.
34 AECL Board of Directors minutes, 27 February 1973.
35 Ibid.
36 AECL Executive Committee minutes, 15 March 1973.

CHAPTER NINE

1 Alec Stewart, interview with author, Kingston, Ontario, 13 october 1989.
2 Boyer, *By the Bomb's Early Light*, 357–8.
3 Lewis, "Canada, the Quality of Life and Nuclear Energy," 12 June 1970, CRNL Records, SDDO, DL-103; Lewis, "Nuclear Energy and the Quality of Life," 1972, AECL 4380.
4 Lewis, "Nuclear Energy and the Quality of Life," 1972, CRNL Records, SDDO, AECL 4380.
5 Lewis, "Nuclear Energy Source for the Future," 26 February 1974.
6 Lewis, "Energy in the Future of Man," 7 October 1975.
7 Lewis's publications on fusion were limited. Lewis, "Controlled H-Fusion," 29 January 1958; Lewis, "Controlled Fusion," 5 May 1971, CRNL Records, SDDO; Lewis, "Energy in the Future – The Role of Nuclear Fission and Fusion," *Proceedings of the Royal Society of Engineers* 70A, no. 20 (1971–72) 219–33.
8 See Rhodes, *The Making of the Atomic Bomb*, 32–5, for a brief discussion of Polanyi's ideas. See Shils, *Criteria for Scientific Development*, for more on the "Republic of Science" and the debate surrounding the criteria of choice for scientific projects.
9 Polanyi, "The Republic of Science," 55.
10 Lewis, "Chairman's Opening Remarks," 12 May 1977, John Foster file.
11 See Weart, *Nuclear Fear*, for a fascinating discussion of the changing attitudes towards both nuclear power and nuclear weapons since the Second World War.
12 Lovell and Hurst, "Wilfrid Bennett Lewis," 501.

CONCLUSION

1 See for example Lewis to Right Reverend Hugh Montefiore, 4 April 1975, Queen's University Archives, Wilfrid Bennett Lewis Papers, Box 4.
2 File "handwritten early (1920s and 1930s) notes from Cambridge," "sermon – first thoughts," n.d., Queen's University Archives, Wilfrid Bennett Lewis Papers, Box 35.

Bibliography

PRIMARY SOURCES

Atomic Energy of Canada Limited, Ottawa, Ontario Records
Chalk River Nuclear Laboratories, Chalk River, Ontario Records
Department of External Affairs, Ottawa, Ontario Records
National Archives of Canada, Ottawa, Ontario
  C.D. Howe Papers
  C.J. Mackenzie Papers
  National Research Council Records
Queen's University Archives, Kingston, Ontario
  W.B. Lewis Papers
Public Records Office, London, UK
  United Kingdom Atomic Energy Authority Records (AB)
    AVIA, Telecommunications Research Establishment Records
    Foreign Office Records (FO)
  Ministry of Aircraft Production Record (AVIA)
Churchill College Archives Centre, Cambridge, UK
  John D. Cockcroft Papers
Royal Signals and Radar Establishment, Malvern, UK
  Telecommunications Research Establishment Records

INTERVIEWS

WITH AUTHOR

William Bennett, 11 January 1989, 23 January 1989
Gilbert Bartholomew, 23 June 1986
Hugh Carmichael, 18 June 1986
Hank Clayton, 18 July 1986

Sir Robert Cockburn, 23 September 1987
Harry Collins, 7 July 1986
L.G. Cook, 19 November 1987
Eugene Critoph, 25 July 1987
Alec Eastwood, 12 August 1987
Margaret Elliott, 5 July 1986
S.W.H.W. Falloon, 5 October 1987
John Foster, 25 April 1988
Ursula Gay, 21 June 1986
Gwynedd Gerry, 8 July 1986
J.L. Gray, 10 July 1986
Geoff Hanna, 2 July 1986
Les Haywood, 15 July 1986
Don Hurst, 20 August 1987
Sir J.C. Kendrew, 5 October 1987
J.W. Knowles, 4 July 1986
George Laurence, 12 July 1986
J.A. Lewis, 22 September 1987
A.M. Marko, 29 July 1987
Gwen Milton, 12 April 1989
Ara Mooradian, 28 July 1987
E.C.W. Perryman, 25 June 1986
E.H. Putley, 14 October 1987
J.A.L. Robertson, 21 July 1987
Harold Smith, 2 May 1988
Alec Stewart, 13 October 1989
Gordon Stewart, 11 July 1986
P.R. Tunnicliffe, 14 July 1986
A.G. Ward, 3 July 1986
Donald Watson, 3 August 1987

WITH R.S. BOTHWELL

George Laurence, 24 July 1985, 16 April 1986, 12 August 1986
John Melvin, 26 September 1985
Colonel Curtis Nelson, 17 April 1985
George Pon, 20 October 1986
Donald Watson, 13 November 1985

SECONDARY SOURCES

Bindon, George, and Sitoo Mukerji. "Canada-India Nuclear Cooperation." *Research Policy* 7 (1978): 220–38.

Bothwell, Robert, Ian Drummond, and J.R. English. *Canada Since 1945*. Toronto: University of Toronto Press, 1981.

Bothwell, Robert. *Nucleus*. Toronto: University of Toronto Press, 1988.

Bowen, E.G. *Radar Days*. Bristol: Adam Hilger, 1987.

Boyer, Paul. *By the Bomb's Early Light*. New York: Pantheon Books, 1985.

Canada. Privy Council Office. Science Secretariat. *The Proposal for an Intense Neutron Generator: Scientific and Economic Evaluation*. Ottawa: Queen's Printer, 1967.

– Senate. Special Committee on Science Policy (Lamontagne Committee). *A Science Policy for Canada*. Vol. 1. Ottawa: Queen's Printer, 1970.

– *Proceedings of the Special Committee on Science Policy*, 30 and 31 October 1968. Ottawa: Queen's Printer, 1968.

Clark, Ronald W. *The Greatest Power on Earth*. New York: Harper & Row, 1980.

– *Tizard*. London: Methuen, 1965.

Crowther, J.G. *The Cavendish Laboratory, 1874–1974*. London: Macmillan, 1974.

Dawson, Frank G. *Nuclear Power*. Seattle: University of Washington Press, 1976.

Doern, G. Bruce. *Science and Politics in Canada*. Montreal: McGill-Queen's University Press, 1972.

Eayrs, James. *Canada in World Affairs, October 1955 to June 1957*. Toronto: Oxford University Press, 1965.

Eggleston, Wilfrid. *Canada's Nuclear Story*. Toronto: Clarke, Irwin & Company, 1965.

– *National Research in Canada*. Toronto: Clarke & Irwin, 1978.

Elliott, L.G., W.B. Lewis and A.G. Ward. "ING: A vehicle to new frontiers." *Science Forum* 1, no.1 (February 1968): 3–7.

English, John. *The Worldly Years: The Life of Lester Pearson Vol. II: 1949–1972*. Toronto: Alfred A. Knopf Canada, 1992.

Goldschmidt, Bertrand. *The Atomic Complex*. La Grange Park, Illinois: American Nuclear Association, 1982.

– *Pionniers de l'atome*. Paris: Stock, 1987.

Gowing, Margaret. *Britain and Atomic Energy*. London: Macmillan, 1964.

– *Independence and Deterrence, 1945–1952*. 2 vols. London: Macmillan, 1974.

Hartcup, G., and T.E. Allibone. *Cockcroft and the Atom*. Bristol: Adam Hilger, 1984.

Hayes, F. Ronald. *The Chaining of Prometheus*. Toronto: University of Toronto Press, 1973.

Hendry, John, ed. *Cambridge Physics in the Thirties*. Bristol: Adam Hilger, 1984.

Hewlett, Richard G., and Oscar E. Anderson, Jr. *A History of the United States Atomic Energy Commission, Vol 1, The New World 1939–1946*. Washington: U.S. Atomic Energy Commission, 1972.

Hewlett, Richard G., and Francis Duncan. *A History of the United States Atomic*

Energy Commission, Vol 2, Atomic Shield, 1947–1952. Washington: U.S. Atomic Energy Commission, 1972.

Hilts, Philip J. *Scientific Temperaments*. New York: Simon & Schuster, 1982.

Jarrell, Richard A. *Cold Light of Dawn*. Toronto: University of Toronto Press, 1988.

Jones, R.V. *Most Secret War*. London: Coronet Books, 1979.

Kevles, Daniel J. *The Physicists*. New York: Vintage Books, 1979.

King, M. Christine. *E.W.R. Steacie and Science in Canada*. Toronto: University of Toronto Press, 1989.

Kinsey, Gordon. *Bawdsey – Birth of the Beam*. Lavenham: Dalton, 1983.

Kulkarni, R.P., and V. Sarma. *Homi Bhabha: Father of Nuclear Science in India*. Bombay: Popular Prakashan, 1969.

Laurence, G.C. "Nuclear Reactor Safety in Canada." *Diversa* (Fall 1988): 16–26.

Lovell, Sir Bernard, and D.G. Hurst. "Wilfrid Bennett Lewis, 1908–1987." *Biographical Memoirs of Fellows of the Royal Society* 34 (1988): 453–509.

MacLeod, Roy M., and E. Kay Andrews. "The Origins of the D.S.I.R.: Reflections on Ideas and Men, 1915–1916." *Public Administration* 48 (Spring 1970): 23–48.

– "Scientific Advice in the War at Sea, 1915–1917: The Board of Invention and Research." *Journal of Contemporary History* 6, no. 2 (1971): 3- 40.

Marko, A.M., G.C. Butler, and D.K. Myers. "Biochemistry at Chalk River." *Bulletin of the Canadian Biochemical Society* 23 (June 1986): 13–19.

McNeill, K.G. "ING: A good thing – under certain conditions." *Science Forum* 1 no. 1, (February 1968): 10–11.

McNelly, M.J. "A Heavy-Water Moderated Power Reactor Employing an Organic Coolant." In United Nations. Conference on the Peaceful Uses of Atomic Energy. *Proceedings of the Second United Nations Conference on the Peaceful Uses of Atomic Energy*, vol. 9, 79–87. Geneva: United Nations, 1958.

Mishra, D.K. *Five Eminent Scientists*. Delhi: Kalyani Publishers, 1976.

National Research Council of Canada. *National Research Council of Canada Review*. Ottawa: NRC, 1948–52.

Oliphant, M. *Rutherford Recollections of the Cambridge Days*. Amsterdam: Elsevier Publishing, 1972.

Parr, J. Gordon. "ING: The wrong thing, in the wrong place, at the wrong time." *Science Forum* 1 no. 1 (February 1968): 8–10.

Patterson, Walter C. *Nuclear Power*. Middlesex: Penguin Books, 1983.

Pilat, Joseph F., Robert E. Pendley, and Charles K. Ebinger, eds. *Atoms For Peace: An Analysis after Thirty Years*. Boulder, Colorado: Westview Press, 1985.

Polanyi, Michael. "The Republic of Science." *Minerva* 1, no. 1 (Autumn 1962): 54–73.

Postan, M.M., D. Hay, and J.D. Scott. *Design and Development of Weapons Studies in Governmental and Industrial Organisation*. London: HMSO and Longmans, Green and Co., 1964.

Pringle, Peter, and James Spigelman. *The Nuclear Barons*. London: Sphere Books, 1983.

Rhodes, Richard. *The Making of the Atomic Bomb*. New York: Simon and Schuster, 1986.

Rigden, John S. *Rabi*. New York: Basic Books, 1987.

Rowe, A.P. *One Story of Radar*. Cambridge: Cambridge University Press, 1948.

Science Council of Canada. *The Proposal for an Intense Neutron Generator: Initial Assessment and Recommendations*. Ottawa: Queen's Printer, 1967.

Sharma, Dhirendra. "India's Lopsided Science." *Bulletin of the Atomic Scientists* 47, no. 4 (May 1991): 32–6.

Shils, Edward. *Criteria for Scientific Development: Public Policy and National Goals*. Cambridge, Massachusetts: MIT Press, 1968.

Shoenberg, David. "W.B. Lewis, 1908–1987." *The Caian* (November 1987): 97–103.

Thistle, Mel. *The Inner Ring*. Toronto: University of Toronto Press, 1966.

Thomson, George P. *J.J. Thomson and the Cavendish Laboratory*. London: Thomas Nelson and Sons, 1964.

United Nations. Conference on the Peaceful Uses of Atomic Energy. *Proceedings of the Second United Nations Conference on the Peaceful Uses of Atomic Energy*. 33 volumes. Geneva: United Nations, 1958.

Wark, Wesley K. *The Ultimate Enemy: British Intelligence and Nazi Germany, 1933–1939*. Ithaca and London: Cornell University Press, 1985.

Watson-Watt, Robert. *The Pulse of Radar*. New York: Dial Press, 1959.

Weart, Spencer R. *Scientists in Power*. Cambridge, Massachusetts: Harvard University Press, 1979.

– *Nuclear Fear*. Cambridge, Massachusetts: Harvard University Press, 1988.

Wedlake, G.E.C. *SOS: The Story of Radio-communication*. Newton: Abbot David & Charles, 1973.

Williamson, Rajkumari, ed., *The Making of Physicists*. Bristol: Adam Hilger, 1987.

Wilson, David. *Rutherford*. Cambridge, Massachusetts: MIT Press, 1983.

Zimmerman, David. *The Great Naval Battle of Ottawa*. Toronto: University of Toronto Press, 1988.

Index